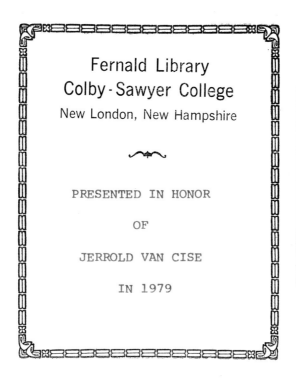

HISTORY, PHILOSOPHY AND SOCIOLOGY OF SCIENCE

Classics, Staples and Precursors

HISTORY, PHILOSOPHY AND SOCIOLOGY OF SCIENCE

Classics, Staples and Precursors

Selected By

YEHUDA ELKANA
ROBERT K. MERTON
ARNOLD THACKRAY
HARRIET ZUCKERMAN

FRENCH SCIENCE

AND ITS PRINCIPAL DISCOVERIES
SINCE THE SEVENTEENTH
CENTURY

BY

MAURICE CAULLERY

ARNO PRESS

A New York Times Company

New York — 1975

Reprint Edition 1975 by Arno Press Inc.

Reprinted from a copy in
  The Princeton University Library

HISTORY, PHILOSOPHY AND SOCIOLOGY OF SCIENCE:
Classics, Staples and Precursors
ISBN for complete set: 0-405-06575-2
See last pages of this volume for titles.

Manufactured in the United States of America

————◆————

Library of Congress Cataloging in Publication Data

Caullery, Maurice Jules Gaston Corneille, 1868-1958.
   French science and its principal discoveries since
the seventeenth century.

   (History, philosophy and sociology of science)
   Translation of La science française depuis le XVIIe
siècle.
   Reprint of the 1934 ed. which was privately printed
in New York for the French Institute.
   Bibliography: p.
   Includes index.
   1. Science--History--France. 2. Scientists--France.
I. Title. II. Series.
Q127.F8C33 1975   509'.44   74-26256
ISBN 0-405-06584-1

FRENCH SCIENCE

FRENCH SCIENCE

AND ITS PRINCIPAL DISCOVERIES
SINCE THE SEVENTEENTH
CENTURY

BY

MAURICE CAULLERY

Membre de l'Institut de France
Professeur à la Sorbonne

La Science est l'œuvre de l'esprit humain, qui est plutôt destiné à étudier qu'à connaître, à chercher qu'à trouver la vérité.

Évariste Galois.

To

Mr. Ormond G. Smith

who inspired it

I dedicate this book as
a token of high esteem
and deep understanding

Foreword

Since the lectures published in this book were heard at the French Institute in the month of March, 1933, Mr. Ormond G. Smith, who had conceived the idea of their presentation, was suddenly called from this life.

The few weeks which I spent in New York afforded me the opportunity to appreciate Mr. Smith's fine qualities of mind, his high ideals, and his disinterested efforts in behalf of that splendid organization of which he was president and which he so ably directed.

Deeply stirred by the death of this noble man, I express here the great sorrow which is mine and which is felt by all those of my countrymen who understood his love of the intellectual achievement and his devotion to the spiritual qualities of France.

I trust that the contents of this book will prove to the American public that Mr. Ormond G. Smith's sentiments were fully justified.

MAURICE CAULLERY.

TABLE OF CONTENTS.

March 3, 1933.

First Lecture

THE CARTESIAN PERIOD OF FRENCH SCIENCE.

(The 17th century and first half of the 18th century.)

———

Some French men of science of the Renaissance.—
The seventeenth century: Descartes.—Le Père
Mersenne and the foundation of the Académie des
Sciences (1666).—The Observatory of Paris and
Astronomy.—The natural sciences.—Réaumur.—
Peyssonnel and the animal nature of coral.

The discoveries of Newton and their influence in
France.—Clairaut, d'Alembert.—Physics in the
eighteenth century.—French society in the eighteenth century and science.

———

The expressions of gratitude and modesty by which
any lecturer comes into touch with his audience,
through constant repetition have become more than
trite; in this instance, however, they are absolutely
necessary and really sincere.

How could I fail to thank this institute, and in a
very special way, its eminent president, Mr. Ormond
G. Smith, for giving me the opportunity to present in
New York, before an audience "d'élite," a comprehensive outline of the great discoveries and major advances
of French science, and to show its creative fecundity?
France has played an essential part in the creation of

modern science and I hope to succeed in convincing you of it.

But, at the same time, how could I help feeling the full weight of the task I am bold enough to undertake? Am I not in duty bound to speak to you of all sciences, mathematics, physics, chemistry, biological sciences, as if none of them, in any of its branches, had any secrets for me? Yet, as a matter of fact, nowadays, the most laborious and gifted men of science are compelled to limit their efforts, not even to the field of one science, but to some more or less restricted corner of it, and I do not pretend to have done anything else. Still, is it not also necessary, at times, to rise beyond the limited horizon of the specialist and as Molière said to acquire "some glimpses of everything," in an effort to ascertain where one stands in the general field of knowledge? To do so, fortunately, one may lean on the achievements of others and look to them for the indispensable framework and guidance. In my case, I was able to find both in recent publications on French science, and, far from hiding my sources, I invoke them to justify my undertaking.

On the occasion of the San Francisco World's Fair in 1915, the French Ministry of Public Instruction had a book of general reference composed on *La Science Française*[1]: each one of its chapters, devoted to a particular science, was written by a specialist of recognized standing. A new edition, entirely revised and brought up to date, is soon to be published in connection with the approaching Chicago Century of Progress Exposition. Moreover, my eminent fellow academician, Gabriel Hanotaux, on the eve of the Great War, had the bright idea to direct on a new plan, the writing of a vast *History of the French Nation*. He did not conceive of it, in the usual fashion, as a series of parts and chapters, each devoted to one of

[1] *La Science Française,* Paris, 1915, 2 volumes.

the successive periods of the unfolding of time, and each treating simultaneously of all the aspects of national activity; instead, he intrusted to a group of qualified specialists the separate history of each aspect of French life through its whole development: *political, diplomatic, economic, military, religious history, history of the arts, history of the sciences, etc.*[1]

I, myself, had the honor of contributing the history of biological sciences, and in the parts of the work dealing with the other sciences, I found, for these lectures, information and views which have given me the most precious help. Nevertheless, I appeal to the forbearance of the men of science who may be in this audience. Each one of them, as I venture in his special field, would be justified in criticizing this outline. To avoid such criticism would be impossible, and, anyhow, the organizers of these lectures had in view an educated public rather than specialists.

*
* *

Science is essentially a universal undertaking, independent of boundaries and nationalities. It is not, however, a frivolous enterprise to follow its advance in a particular country: first, one may legitimately aim —as it is precisely your intention—to find out what contributions a certain country has made to scientific progress; and further, from a more general point of view, in each country, on account of the particular traits of its civilization and culture, scientific progress assumes a specific character and pace. It is interesting to analyze what elements have favored, therein, or

---

[1] Gabriel Hanotaux—*Histoire de la Nation Française,* 15 volumes in 4° (librairie Plon.) *L'Histoire des Sciences* forms the last two volumes. *Introduction à l'Histoire des Sciences* by Emile Picard; *Histoire des Sciences Mathématiques et de l'Astronomie,* by H. Andoyer and P. Humbert; *Histoire de la Physique* by Ch. Fabry; *Histoire de la Chimie* by A. Colson; *Histoire des Sciences Biologiques* by M. Caullery.

hindered the rise of science, what part have played institutions and establishments peculiar to that country. Each nation also has its own mentality, which undeniably exerts an influence on its scientific life. Thus, it appears to me, French science has highly individualized characteristics which I shall try to emphasize for you. This survey would then find further justification apart from your kind interest in French productions.

Needless to say, the very limited time at my disposal forbids my attempting detailed accounts, or striving for completeness, and even more bringing you such a wealth of information as can be found in the two works I mentioned a while ago. Out of the mass of French scientific production, my intention is to limit myself to a choice of capital achievements, to throw the light on the great men of science, whose new thoughts exerted a commanding influence in and outside of France, upon their own or succeeding generations. In addition, I shall trace the influence upon them of the scientific masterpieces of foreign origin. In lectures of this kind, it is not the least difficulty to choose judiciously what is of real significance without omitting any essentials, and to convey an idea of it, as briefly as possible, avoiding at the same time dryness and obscurity.

I had to choose between two main plans. I could either take up successively each of the sciences, examining what contributions should be credited to French discoveries or trace simultaneously the progress of the various sciences during separate periods. Upon consideration, I decided in favor of this second way of presentation. The perspective it affords, appears, I believe, both more vivid and coherent. In any period, the various sciences have reacted upon one another. The minds of the scientists, the aim of their preoccupations and efforts closely depend on the whole body of contemporary knowledge. Neither the men nor the works can be separated from their time. It is hardly possible to appreciate fully the nature and importance of discoveries unless we consider them in relation to the period in which they are made. Such admiration as some of them should arouse in us increases in so far as they appear to be well ahead of their time. Therefore, it seemed to me that the subject you asked me to treat would acquire real unity and proper relief if treated by periods.

I have, therefore, adopted a historical point of view. Except for a few very brief indications on the previous period, I shall begin with the seventeenth century; not because France lacked men of science before then, but because it was in the seventeenth century that scientific life was organized and soared away. It was from that time that fundamental discoveries were made in France, and that here, as well as elsewhere, the true foundations of modern science were laid down. And as your interest lies in modern science, I do not propose to study the archeology of scientific thought, but to rediscover the vital sources of present-day science.

I am thus led to divide my outline as follows:

1.—To-day, I will take you rapidly through the seventeenth century and the first half of the eighteenth century.

2.—The second lecture will be devoted to the second half of the eighteenth century.

3.—The third will cover the beginning, or more precisely, the first three decades (1800-1830) of the nineteenth century, a period which has been, perhaps, the most brilliant in the history of French science.

4.—The fourth will deal with mathematical sciences, physics and chemistry during the remainder of the nineteenth century (1830-1900).

5.—The fifth will study biological sciences during the same period.

6.—The sixth, finally, will take up contemporary

science, the science the men of my generation have "lived," during the last thirty or forty years, and will end by drawing up general conclusions.

Of course for none of the above periods is there any question of rigid limits. In this field as in all others, life is essentially continuous and the various events overlap each other. The limits assigned can be nothing but approximate.

*

* *

Men of Science in the Renaissance (sixteenth century).

The sixteenth century saw the blossoming of a long movement, the Renaissance, the origins of which go back to the thirteenth century; it was mostly a reëstablishment of contact with the civilization of antiquity and a slow reassimilation of its intellectual treasures. Italy heralded this great movement which later on extended to the North, in France and in England. The feeling of wonder thus aroused in eminent minds generally led to a blind admiration of antiquity. On all matters, it seemed that truth could be found only in conformity with what the ancients had upheld. In a large measure, the modern spirit has consisted in an emancipation from the fetichism of this authority and in a tendency to scrutinize Nature without prejudice. This spirit has met with much resistance. Each of its great conquests has had to contend on one side with uncritical admiration of antiquity, on the other with such perturbations as the new ideas caused in traditions and religious beliefs. The emancipation of thought begins in the seventeenth century and this is the reason for choosing it as our starting point. In 1633, Galileo still incurred condemnation from the Inquisition on account of his demonstration that the earth moves round the sun. This was not to be the last obstacle raised by

the spirit of authority on the path of scientific progress. We shall see that in the eighteenth century, the
Sorbonne in the same manner condemned the *Théorie
de la Terre* of Buffon who, just like Galileo, yielded,
at least outwardly. Nay, did we not ourselves in recent
years, much to our surprise, witness in the United
States a belated outbreak of the old spirit forbidding
the teaching of Evolution in some States of the Union?

Conditions under which scientific research labored
in France at the beginning of the seventeenth century
had nothing in common with those now prevailing. In
Italy the universities, even before the sixteenth century, had been seats of fruitful thinking, which in
particular had renovated anatomy and where men of
Galileo's type taught; in France, on the contrary, the
universities, ever since the thirteenth century, had been
submerged by the teaching of Scholastic Philosophy
and Theology. Alone, the University of Montpellier,
owing to its location and immediate contact with the
Mediterranean world, perhaps, also, to its partly Arabic
origins, was able to escape in a measure the steriliz-
ing domination of medieval thought; it thus remained
until the French Revolution of 1789, at least in medicine and biological sciences, an independent and active
center. The University of Paris, on the other hand,
one of the first to be established in Europe, and ever
since looked upon as a model, had in the thirteenth century drawn crowds of foreign students eager for learning; but it had more than any other borne the yoke
of Scholasticism and Theology. Rabelais's satire is
largely devoted to the arraignment of this mentality.
It was as a reaction against this spirit of obscurantism
that in the sixteenth century, in 1530, King Francis I
instituted Royal Lecturers face-to-face with the university; they formed, later on, the Collège Royal, our
present Collège de France, a center of humanism and
modern spirit. To realize better this feeling of antago-

nism, let us listen to a passage of an oration Pierre de la Ramée (Ramus), one of the lecturers of the Collège Royal, addressed to King Charles IX around 1565; the author rebels against the scholastic tendencies in medical teaching as given in Paris. He requests that, for some part of the year, students be taken by their master through fields and forests and therein taught to "philosophize" over herbs and plants; during another part of the year they should be trained in dissecting bodies; and in another part of the year they should, above all, fully participate in consultations and in treating the sick. "Cette méthode," adds Ramus, "ferait des médecins comme elle en fait à Montpellier et dans toutes les écoles de médecine d'Italie. La dispute scolastique peut seulement produire des scolastiques, mais non des hommes guérissant les maladies. . . . Réalisez," he says to the king, "ces méthodes, afin de produire non seulement des érudits sur les théories des livres, mais des hommes formés à la guérison des malades par les exemples et la pratique."[1]　A program indeed fully modern in spirit and scientific in the present sense of the word. Still, the theological spirit was to dominate in the universities up to the time of the Revolution, and in France, prior to the nineteenth century, we need not look to them in our search for the sources of scientific progress. This is one of the principal features of the history of science in France. With us, science from the first, and for a long time, was the spontaneous creation of isolated individuals who, aside from their avocations, dedicated their leisure to it with a pas-

---

[1] "This method," adds Ramus, "would train physicians in the same way they do at Montpellier and in all the schools of medicine of Italy. Scholastic disputations can but produce school men, not in any way men able to cure diseases. . . . Bring such methods into practice," he says to the king, "so as to produce not only erudites on the theories of books but men trained in curing the sick through examples and practice."

sionate interest. Most of the leading men of science were, for example, physicians, magistrates, or councillors of provincial Parliaments. In spite of such unfavorable conditions, France, as early as the sixteenth century, gave birth to illustrious men of science. I will recall the names of a few:

François Viète (1540-1604), a counselor at law and a magistrate, played a part in public affairs of importance, became a member of the king's privy council under Henry IV. He is looked upon as the greatest mathematician of the sixteenth century. The founder of modern algebra, the creator of trigonometry, he has left just as deep an imprint on the advance of geometry. His reputation was great in his time, but later on he was unjustly forgotten and remained so for a long time.

Another very great mind and admirable moral character was Bernard Palissy, a self-taught man, who, apart from any tradition, had intuitions of important truths: "Je ne suis," disait-il, "ne grec, ne hébreu, ne poète, ne rhétoricien, ains un pauvre artisan bien pauvrement instruit aux lettres."[1] He owes most of his lasting fame to his skill in the art of ceramics; but he cogitated on a mass of subjects and set down a number of original and judicious ideas in small treatises; these are written in a French of a charming "naïveté" notably in the form of dialogues between two symbolic characters, Théorique and Pratique. Théorique embodies the School, scholasticism, generally accepted ideas; Pratique, the free observation of Nature. Pratique in no way symbolizes routine; on the contrary, she believes in the power of method. B. Palissy had, for his time, extremely penetrating views on agriculture, physics and chemistry; he studied directly marine animals on the coasts of Saintonge;

---

[1] "I am," said he, "neither Greek nor Hebrew, neither a poet nor a rhetorician, but a poor artisan very poorly taught in letters."

but, above all, he was a real forerunner in geology, and his name should survive as that of one of the pioneers of modern science.

Likewise, it would be appropriate to recall among his contemporaries, at least Guillaume Rondelet (1507-1566), a professor of anatomy in the Faculty of Montpellier, and Pierre Belon (1517-1565), both zoölogists of real merit; the great surgeon Ambroise Paré (1509-1590), as well as several botanists whose center of activities was mostly Montpellier, where a great botanical garden was established in 1593, patterned after those already in existence in Italy. So we see that there were in France, in the sixteenth century, rather numerous scientific personalities working, more or less, in isolation, either self-taught or developed under the influence of Italy.

*
* *

The Seventeenth Century—Descartes.

One must reach the seventeenth century to see French thought become, in its turn, an inspiring force which radiated far and wide. In the foremost rank of these initiators, one must place Descartes. His name sums up the spirit of all science for a century.

Descartes (1596-1650), even to the present day, remains one of the great classical philosophers; he personifies the emancipation of modern thought from scholasticism and medievalism. He created the conditions which were essential for the birth and the development of scientific thought. But he is at the same time one of the great creators of modern science itself, even though on this ground his works retain an interest only in retrospect. At any rate, accepted or disputed, his thought dominated almost unrivalled the seventeenth century and the first part of the eighteenth century until the time when it yielded ground slowly to Newton's ideas. Therein lies the unity of the period which we are studying to-day, and which may truly be called the period of Cartesian Science.

Nor was René Descartes a professional man of science. Educated at the Jesuit "Collège" at La Flèche, he had started on a military career at the time of the Thirty Years' War; then forsaking the army, he spent several years in free meditation, getting rid of the influence of any school.

As early as 1629, he took up his residence in Holland, then the land of freedom of thought. Even there he did not feel completely secure, since after Galileo's condemnation he destroyed, from caution, the *Système du Monde* which he had composed. His *Discours de la Méthode,* published in 1636, is the manifesto of modern thought. Therein he overthrows the system

René Descartes

built up by scholasticism, driving away all the verbal phantoms the schoolmen had conjured. Razing to the ground, as a preliminary, this whole structure, he raises a new one, from the corner stone up, establishing, to do so, safe and unassailable rules of reasoning which constitute the method to be followed by thought. From this should derive all science, and Descartes, trusting to this guidance with the boldness of a genius who holds no doubt about his powers, undertakes the building up of universal science, founding it as much as possible on the observation of reality; but in his haste, he could not wait until observation had furnished him with materials and deductive reasoning was called upon to bridge over the gaps.

The *Discours de la Méthode,* moreover, included three appendixes: *Traité de la Dioptrique, Traité des Météores, La Géométrie,* which were simply illustrations of the method applied to particular cases. Now, *La Géométrie* was a capital contribution to the advance of mathematics, brimful with new data and ideas. For the first time, Descartes utilizes the resources of algebra, becoming the creator of analytic geometry, which extended prodigiously the field of geometrical science. Descartes also made very valuable contributions to algebra. In mechanics, he initiated large and sound points of view, notably on the principle of inertia and on the fundamental value of the notion of force. His *Dioptrique* in which he applies the data of geometry is a considerable piece of work, most helpful to the advance of optics, particularly for the study of the refraction of light; he set down in it fundamental problems, even though he did not solve them accurately. He deduced from it the theory of optical instruments and in particular the theory of vision; he had the intuition of betterments which were not achieved until much later and, besides, according to different principles. Descartes went as far as giving a complete explanation of the Universe, reducing everything to motion, that is to say, to mechanics and mathematical calculation. The world is nothing but a system of vortices; the same is true even of the human body and functional activity of living creatures. It is on these ideas that the seventeenth century was to live. Undoubtedly, such Cartesian conceptions have since yielded to others, which in their turn have been more than once renewed; but is it not an awe-inspiring contribution to science to have substituted for the old ideas a system of thought embracing the universe in its entirety, a system, which after ruling as a whole, for a century, still left deep marks and many definite results?

*
* *

Le Père Mersenne and the Foundation of the Académie des Sciences (1666).

The immediate diffusion of the works and thoughts of Descartes leads us to inquire into the conditions under which, in his time, scientific ideas and results were being spread. Outside of books, the editing of which was rather difficult on account of strict censorship and heavy condemnations—Holland enjoying in this regard a certain degree of freedom—there was no regular and open way for men of science to communicate with one another. The difficulties attached to the publishing of books in the seventeenth and eighteenth centuries account for the fact that the publication of capital works was often long delayed, sometimes even taking place only at more or less great intervals after the death of their authors. There existed in the time of Descartes neither scientific journals nor established academies or learned societies.

This is the reason why men of science had to rely on letters to each other as a means of disseminating scientific news; it was also the custom for a scientist, after elucidating or solving a problem, to challenge his correspondents to solve it, and in many cases the challenger's solution has not reached us. In this intellectual intercourse during the first half of the seventeenth century, a man played an important part without being himself a scientist of the first rank. Father Marin Mersenne (1588-1648). A Minimite friar and a friend of Descartes, who, like him, had been a former pupil of the Jesuit "Collège" at La Flèche, Mersenne multiplied the contacts between his scientific co-workers, giving information to all of them, instigating their communications, taking trips here and there.

Thus, for instance, did he meet Galileo, acquainted France with the famous experiment of Torricelli, on the rising of the mercury column in a vacuum tube. He edited works of various men of science. Lastly, he organized in Paris more or less regular meetings between amateurs and his friends practicing science; it is from these meetings, continued after his death, that resulted the foundation of the *Académie des Sciences* in 1666.

Let us dwell a while on this group of scientists who gravitated around Mersenne, since some of them were men of the very first order.

Among his correspondents, one must mention first Pierre de Fermat (1601-1665), who engaged with Descartes in spirited controversies and was often on the right side. Here again was an amateur, an amateur of genius. Fermat, like Viète, was a magistrate, a counselor of the Parliament of Toulouse, and does not seem to have ever left the south of France. In his time, and rightly so, he was looked upon as the greatest geometrician in Europe, yet he published very little. His works and letters were collected only after his death by his son, at the cost of great difficulty and unfortunately in a very incomplete manner. We do not possess by far all he had found. He was content, most of the time, to write down his findings merely as marginal notes on the works of others, without taking the trouble of setting down the demonstration. Such was the case of the famous theorem commonly known, even to-day as Fermat's theorem[1]; he claimed to have proved it, but did not tell how, and a general demonstration of it still remains to be found. Fermat excelled above all in the theory of numbers, one of the

---

[1] No integral values of a, b, c can be found to satisfy the equation $a^m + b^m = c^m$, if m is an integer greater than 2.

most abstract and highest branches of mathematics, which, up to our day, has had a particular attraction for the greatest geometricians; he also excelled in geometry. He certainly was one of the greatest mathematical geniuses of modern times. In what we have of his works, in regard to problems of maxima and minima, we find methods of calculation which are really the basis of what was to be infinitesimal calculus; a few decades later, Newton and Leibnitz both fully worked out infinitesimal calculus, kept it a secret for a long while, and both bitterly claimed priority of invention. Now, Leibnitz had visited Paris and might have been aware of the results obtained by Fermat. Fermat must also be regarded as one of the inventors of the mathematical theory of probabilities.

Descartes and Fermat occasionally clashed; the latter was as conciliatory as the former was pugnacious; more than once Fermat, not Descartes, was right. Such was the case in regard to the refraction of light. Descartes' ideas on the propagation of light wavered; on one side, he conceived of it as instantaneous, on the other, he claimed that it had greater speed in water than in air. Fermat, on the basis of considerations about the minimum of time, supported the opposite view and the future acknowledged he was right; experimental proof could not be brought forth until Foucault's work in the nineteenth century.

In the group of Mersenne, we find another magistrate, Etienne Pascal (+ 1651), also a notable scientist, whose name, however, survives mostly through his son, Blaise Pascal, equally renowned in literature and in science. Blaise Pascal (1623-1662), a child of genius, whose brief career ended in mysticism, had, as early as the age of sixteen, already made remarkable discoveries in geometry and set them down in his *Essai pour les Coniques*. At the age of nineteen, he invented his computing machine to relieve his father whose work required very long calculations. In the course of a few years, he kept adding up important discoveries and expressing most profound ideas, bringing forth the initial fundamental contributions to the theory of probability and integral calculus. At the same time, he was a very great physicist. He realized immediately the far-reaching import of Torricelli's experiment, saw in it proof that air is a substance having weight, and to verify it, actually, had his brother-in-law, Périer, conduct the famous experiment of the Puy-de-Dôme. It consisted in comparing the height of the mercury column in the Torricellian tube at the top of a mountain and at Clermont-Ferrand. The barometer was the result of this experiment. If one wishes to gauge the deep and far-reaching significance of these experiments, from the point of view of the modern mind, it is enough to recall this conclusion of the *Traité sur le vide et la pesanteur de l'air:*

Blaise Pascal

"Que tous les disciples d'Aristotle assemblent tout ce qu'il y a de plus fort dans les écrits de leur maître et de ses commentateurs pour rendre raison de ces choses par l'horreur du vide s'ils le peuvent; sinon qu'ils reconnaissent que les expériences sont les véritables maîtres qu'il faut suivre dans la physique."[1]

---

[1] "Let all the disciples of Aristotle get together all that is strongest in the writings of their master and let his commentators give an account of these things by the abhorrence of a vacuum, if they can; if not, let them acknowledge that experiments are the only real masters one must follow in physics."

In his *Traité de l'équilibre des liqueurs,* he conceived and formulated the fundamental principle of hydrostatics in reference to pressures exerted by fluids on the sides of vessels and derived from it the hydraulic press. This principle is named for him.

By the side of these men of genius, one should also mention, among the correspondents and intimates of Mersenne, Girard Desargues (1593-1662), an engineer and architect, one of the creators of modern geometry; G. Personne de Roberval (1602-1675); Fabri de Peiresc (1580-1637), an astronomer, counselor of the Parliament of Aix; Gassendi (1592-1655), an astronomer and philosopher; and others. I cannot dwell further on this already distant period, and what I have said is enough to show the exceptional value of this group of French men of science in the middle of the seventeenth century; it numbered many remarkable individuals, most of whom cultivated science as a side line to their regular avocations, and among whom we find some of the principal founders of modern mathematics and physics.

Father Mersenne, did we say, served as a connecting link between men of science, some of whom lived in Paris, others in the provinces, others in foreign countries. After his death the meetings that he had instituted were continued in the houses, at first, of the maître des requêtes de Montmort, later of Melchissédec Thevénot, a diplomat and a friend of science. From these gatherings, which Chancellor Francis Bacon mentions, finally evolved the *Académie des Sciences;* it was organized by the great minister Colbert in 1666, at the instigation of Claude Perrault, the architect of the colonnade of the Louvre and a distinguished scientist. Under similar conditions had Cardinal de Richelieu founded in 1635 the French Academy, taking as a nucleus meetings of men of letters held at the home of one of them, Valentin

Conrart.　Quite in the same manner, by the way, the Royal Society of London was founded in 1660 as a successor to private meetings held at Oxford for some years by a few men of science.

*

*　*

The Paris Observatory and Astronomy.

At the same time he founded the *Académie,* Colbert suggested to King Louis XIV the construction of the Observatory of Paris (1667-1672), to make easier astronomical researches and provide laboratories for the academicians.　Cl. Perrault was the architect of this building, which is still used for the same purpose.　France had already had eminent astronomers, Fabri de Peiresc, Gassendi, both already mentioned, l'abbé Picard (1620-1682), the author of astronomical and geodesic works of the first rank, and Auzout (1630-1691); they contributed efficiently to progress, perfecting methods of observation and instruments. Besides, eminent foreign astronomers were called to the new observatory; they increased the vitality and brilliancy of the new institution.　Among these, the great Dutch scientist, Christian Huygens (1629-1695), who lived in Paris from 1666 to 1681 and might never have left but for the deplorable religious policy of Louis XIV; the climax of this policy, in 1685, was the revocation of l'édit de Nantes, which drove away from France a Huguenot élite, a loss ever to be regretted.　The case of the Danish astronomer Roemer (1644-1710) was quite similar.　Picard had brought him to Paris on returning from Copenhagen, where he had gone to ascertain the location of Tycho-Brahe's observatory.　From 1672 to 1679, Roemer remained at the Observatory of Paris where he made essential discoveries.　The new observatory was made famous by the Italian, Jean-Dominique Cassini

(1625-1712); he settled permanently in France and was the ancestor of a dynasty of astronomers whose glorious succession extended through four generations, over nearly two centuries, up to 1845. Finally, in 1665, was founded the *Journal des Savants* which through various vicissitudes, has been able to survive until the present time. In this manner was scientific life organized and provided with its essential instrumentalities around 1666.

Thanks to these institutions and this concourse of remarkable men, discoveries followed rapidly on top of one another, particularly in the field of astronomy. Cassini made great advances in the knowledge of planets and their satellites, especially those of Jupiter; his observations on the eclipses of the satellites of this planet led Roemer, in 1676, to explain their discrepancies by the different time light required to reach us, according to the distance between Jupiter and the Earth. From this remark, with the intuition of a genius, Roemer deduced the measurement of the speed of light, a capital discovery which was to play an increasingly important part in physics. At the same time, Huygens became famous due to discoveries relative to the satellites and the ring of Saturn, as well as by mathematical works. Above all, did he achieve fundamental results in optics. He found and explained the double refraction of light in Iceland spar and worked out the wave theory of light. This admirable theory is expounded in his *Traité de la lumière* composed in Paris, and communicated to the *Académie des Sciences* in 1678, but which Huygens published only in 1690, after his return to Holland. Starting from the conceptions of Descartes and Fermat, completing them with new discoveries, Huygens conceived light as due to the propagation of a vibration in a medium, and built up the undulatory theory; after being discarded for a century in favor of Newton's theory of emission, the undulatory theory was to receive in the nineteenth century a brilliant confirmation, after Young's and Fresnel's discoveries.

The men of science connected with the Paris Observatory, Cassini, Roemer, and Picard, in 1677, in the neighborhood of Paris, also took the first measurement of the speed of sound; the figure they arrived at was found later to disagree with the one Newton reached a little later by calculation; this discrepancy was only explained in the nineteenth century on the basis of considerations in thermodynamics.

Finally, it is also from the Paris Observatory that originated the quite novel and magnificent project of establishing the measurements of the earth. In 1669 and 1670, the astronomer Picard measured for the first time a segment of the earth's meridian by a triangulation between Amiens and Malvoisine, a locality in the south of Paris; he followed in carrying out his work the method of Snellius, a Dutchman. Newton relied on the measurement of Picard to calculate the force which keeps the moon revolving on her orbit. He found it to be gravitation.

As early as that time, moreover, French astronomers did not limit themselves to taking measurements in France; in 1671, one of them, Jean Richer (1630-1696), was sent to Cayenne, in Guiana, to measure the parallaxes of the Sun, the Moon, Venus and Mars by combining these observations with measurements taken in Paris simultaneously. Whereupon, (Huygens having just invented the pendulum clock) Richer took to Cayenne a pendulum striking the second, Paris time; the unexpected and capital observation was made that the swing was shorter in Cayenne than in Paris. From this fact, and from considerations on centrifugal force, Huygens drew the conclusion that the earth is flattened around the poles; Newton was led to accept this view in some other way. The contrary opinion, that of an elongation in the polar regions having been upheld, the *Académie,* around 1730, was brought to undertake a direct verification of the facts. Two missions were organized for the purpose of measuring the meridian, the first in the vicinity of the pole, in Lapland, the other near the equator, in Peru. The first was under the direction of Maupertuis (1698-1759), the second composed of Bouguer (1698-1758), La Condamine (1701-1774) and Godin (1704-1760). The results obtained concurred with the opinion held by Huygens and Newton.

In all I said so far, I could not attempt to draw a detailed picture of French science in the seventeenth century. I was bound—and from now on will be even more so—to limit myself to facts and men of science of the first rank; still one has been able to see the extent and magnitude of the problems under examination. Many other names of French scientists of that period would deserve recognition besides those mentioned: among the mathematicians, Rolle (1652-1716), whose name is linked with a fundamental theorem of the theory of equations; the Marquis de l'Hôpital (1661-1704); Varignon (1654-1722); in physics, the Abbé Mariotte (1620-1684) who along lines parallel to Rob. Boyle of England, but independent of him, established the laws of the compressibility of air and formulated them in his *Essai sur la nature de l'air* (1679); Denis Papin, a collaborator of Huygens at the observatory, and the first of the forerunners in the invention of the steam engine. The attacks on the protestants drove him away from France, and he lived for a long time in London, became a Fellow of the Royal Society and worked with Boyle.

*

* *

The Natural Sciences.

There has been little mention up to now of natural sciences. In regard to them, it behooves me to recall the foundation in Paris in 1626, under Louis XIII of the *Jardin des Plantes Médicinales* or *Jardin du Roi*, modeled along the lines of the one at Montpellier. Alongside of the study of botany, professorships in anatomy and chemistry were quickly created; in the seventeenth and, even more, in the eighteenth century, this establishment, with the Collège Royal, formed the center of higher scientific learning in Paris. Among the botanists who contributed to its fame, I may mention its first superintendent, Guy de la Brosse (+1641), Fagon (1638-1768), a very receptive mind, physician to Louis XIV, and especially, Joseph Pitton de Tournefort (1656-1708) who ascertained definitely most of the genera of plants. The tradition thus created was kept brilliantly alive all through the eighteenth century by the Jussieus; of the two brothers, Antoine (1686-1758) and Bernard (1699-1777) and of their nephew, Antoine-Laurent (1748-1836), we may say, anticipating on the next period, that they set forth in all clarity the individual character and relationships of the major groups in the vegetal kingdom: families, orders, etc.

In the field of anatomy, so brilliantly investigated in the sixteenth century by the Italian universities, the seventeenth century witnessed the end of the ruling of Galen's ideas which had maintained themselves over fourteen centuries. Their overthrow is a long story, particularly rich in suggestions. Their downfall was finally achieved by William Harvey's discovery of the circulation of blood, which he published in 1628. The Faculty of Medicine of Paris strenuously opposed this novel truth, earning fully, thereby, the sarcasms of Molière. It is to the credit of Descartes that he aligned himself from the very first with Harvey's supporters. An important supplement to the new doctrine was brought forward by a young physician of Montpellier, Jean Pecquet (1622-1674). In 1647, he ascertained the relations between the lymphatic and blood vessels, incurring, on this account, the anathema of the anatomist Riolan[1] and the last upholders of Galen. The liver, until then regarded as the all-powerful center of organic functions, was shorn of its glory. Physiology, though, was still to wait more than a century to make a real start. It consisted, then, merely of hypothetical conceptions of which nothing remains. In physiology, we again find the imprint of Descartes, who took a keen interest in anatomy, and, in his attempt at universal science, applied to living organisms his general mechanistic conceptions. Outworn and archaic as may be his views in physiology, they are not without some interest, for they put him in the ranks of the opponents of vitalism; ani-

---

[1] *"Aussi voyons-nous que 'l'ancienne et véritable médecine' confirmée par l'expérience de tant de siècles, se corrompt et pervertit entièrement, tant par l'introduction de nouveaux monstres d'opinions chimériques que par l'exhibition de mille sortes de médicaments vénéneux pour tuer les hommes impunément. Pecquet a fait bien davantage; il a commencé à bouleverser la structure et composition du corps humain par la doctrine nouvelle et inouïe qui renverse entièrement la médecine ancienne et moderne, la nôtre, tant en la physiologie qu'en la pathologie et thérapeutique."* J. Riolan, *Manuel anatomique et Pathologique, Paris, 1653—Discours contre la nouvelle doctrine des veines lactées (p. 678-695).**

* "So do we witness that *ancient and true medicine* confirmed by the experience of so many centuries, is being corrupted and completely perverted as much by the introduction of novel, monstrous and chimerical opinions as by the public exhibition of thousands of kinds of poisonous remedies to kill people with impunity. Pecquet has done much worse, he has set his hand to turn topsy turvy, the structure and composition of the human body by means of that new, unheard of doctrine, which overthrows completely that ancient and modern medicine we claim for our own in physiology as well as in pathology and therapeutics." J. Riolan, *Manuel anatomique et pathologique, Paris, 1653—Discours contre la nouvelle doctrine des veines lactées (p. 678-695).*

mals, according to him, are merely machines, just like other machines; this view was to be shared all through the century; we find traces of it in La Fontaine; even the devout Malebranche held it, conceiving, it is true, the soul as radically distinct from the body. According to this idea, vital functions are to be explained in terms of space and motion, to the exclusion of any occult forces and scholastic qualities. Physiology is a part of physics and as such is nothing but geometry and mechanics.

"Si l'on connaissait," dit Descartes, "qu'elles sont toutes les parties de la semence de quelque animal, en particulier de l'homme, on pourrait déduire de cela seul, par des raisons entièrement mathématiques et certaines, toute la figure et la conformité de ses membres."[1]

Such specific explanations as Descartes imagined in the absence of any experimental basis, appear to us, naturally enough, very coarse and, to be sure, Pascal very wisely restricted this limitless mechanism.

"Il faut dire, en gros, cela se fait par figure et mouvement, car cela est vrai, mais de dire quels et composer la machine, cela est ridicule, car cela est inutile, incertain et pénible."[2]

L'Académie des Sciences which, ever since its foundation, numbered anatomists and botanists, undertook collective inquiries on the structure and functions of animals and plants, but they did not lead to notable discoveries. Therefore, we will not dwell on

---

[1] "If one knew," said Descartes, "what are all the parts of the germ of an animal, in particular of man, one could deduce from this alone, by exclusively mathematical and certain reasoning, the whole figure and conformity of its members."

[2] "One must say, grosso modo, this is done by figure and motion, for this is true; but to say of what kind and try to combine the machine, this is ridiculous, for this is useless, uncertain and painful."

them; let us pass, immediately, to one of the great figures of French science, who, though belonging to the eighteenth century, may properly be linked with the period now under review, as he remained to the end a faithful believer of Cartesian ideas: Réaumur (1686-1757). Th. Huxley goes as far as to look upon him as the greatest naturalist the world has known between Aristotle and Darwin.

Réaumur.

Réaumur, on many counts, in the first part of the eighteenth century appears to us a scientist of a singularly modern complexion; his work has aged far less than that of his contemporaries. First of all, he was a positive mind, scrupulously devoted to observation and experience, unconcerned with any system or metaphysics; on the other hand he remained respectful of tradition and religion. He was not satisfied with mere observation; he conducted experiments with perfect freedom of mind and great ingenuity. He ranks among the initiators of the difficult art of biological experimentation. Armed with strong mathematical training, not only was he a naturalist, but also, and maybe above all, a great physicist and engineer in the modern sense of the word. At the beginning of his career, he worked out a number of important industrial inventions: the manufacturing of steel, tin plate, and porcelain; a study of the resistance of cables; the perfecting of the thermometer (his graduation of this instrument is still widely used). Rewarded for these immensely valuable contributions by a generous pension from the king, he devoted his life to the study of Nature, without any thought of personal profit. At a very early age, he was admitted to the *Académie* where he exerted a considerable influence, published a long series of works on various subjects; at the same time he guided and inspired numerous correspondents in the provinces and abroad; several of those were themselves illustrious, such as Charles Bonnet of Geneva, Abraham Trembley and Pierre Lyonnet of The Hague, Ch. de Geer of Stockholm; many others showed Réaumur their most affectionate esteem and warmest admiration. It would be much too long to try to analyze, even briefly, the works of Réaumur. As a naturalist, he was noted for his abil-

Réaumur

ity and ingenuity in tracing the animal through all the phases of its living state, observing its habits and relations with the environment. Réaumur was the founder and master of a branch of biology which in our time has been called *ethology* or *œcology*. He made important studies on most groups of invertebrates, and on various general problems, like regeneration or embryology. Of his works, the one which remains the best known and most used is his *Histoire des Insectes* in six volumes; it is unfortunately incomplete. Mr. W. M. Wheeler, the eminent entomologist of Harvard University, had the good fortune, in 1925, to reveal and edit an unpublished memoir which was to be a part of the seventh volume; considering its date, this is a magnificent piece of work on ants. Réaumur, as on many other points, was far ahead of his time.

He held the foremost rank among his contemporaries, and in addition to his own findings, suggested and helped to bring forth many an important discovery.

Such was the case for the capital fact of parthenogenesis which Charles Bonnet, at the age of twenty, derived from a study of plant lice with the help of Réaumur's advice. His very voluminous correspondence, the editing of which would be desirable, shows with what courtesy, disinterestedness and tactfulness he followed the efforts of his correspondents through the whole world. He is, in the full meaning of the word, a great figure.

Peyssonnel and the Animal Nature of Coral.

But, as no one is ever fully safe from error, in one instance Réaumur, through excessive caution hindered the career of a naturalist of real merit; thus, he temporarily suppressed an important discovery: the animal character of coral, believed until that time to be a plant. Antoine Peyssonnel (1694-175?), a naturalist from Marseille, in 1725, made this discovery, in a clear-cut manner, as shown by the following description:

"Je vis fleurir le corail dans des vases pleins d'eau de mer et j'observai que ce que nous croyons être la fleur de cette prétendue plante n'est, au vrai, qu'un insecte, semblable à une petite ortie (actinie) ou poulpe (polype). . . . J'avais le plaisir de voir remuer les pattes ou pieds de cette ortie et, ayant mis les vases où était le corail à une douce chaleur, auprès du feu, tous les petits insectes s'épanouirent. . . . Le calice de la prétendue fleur est le corps même de l'animal avancé et sorti hors de sa cellule."[1]

Réaumur realized his mistake after Trembley's ex-

---

[1] "I saw coral blossom in vases full of salt water, and observed that what we believe to be the flower of this so-called plant is truly nothing else than an insect similar to a tiny sea-nettle (actinie) or polypus. . . . I had the pleasure of seeing the legs or feet of this sea-nettle move and, as I exposed the vases in which the coral stood, to a moderate heat, near the fire, all the little insects began to expand. . . . The calyx of the so-called plant is the body itself of the animal fully out of its cell."

periments, in 1740, on fresh-water hydra. Peyssonnel's bad luck, moreover, did not end there, as he had to wait until 1750 before publishing his remarkable findings in the *Philosophical Transactions of the Royal Society of London.* Peyssonnel did not give his full measure.

*

* *

The Discoveries of Newton and Their Influence in France.

The whole period we have just gone over shows real unity, due to the deep influence of the ideas of Descartes. To all men of that time, the universe appeared as a system of vortices and Descartes's conceptions permeated all science. Antiquated as they may be to-day, they still cut a great figure, and after all, are they so far from our present ideas on the atomic structure of the universe?[1]

---

[1] The analogy I suggest here reveals itself in many peculiar coincidences of expression. I called attention, a few years ago, to a curious passage in the *Histoire des Insectes* of *Réaumur* (T. II, *Préface,* p. XXXI-XXXII) which sounds like an anticipation of present-day views on the structure of the atom. Here are the essentials of this passage of pure Cartesian inspiration as the rest shows: Réaumur imagines that one finds: *"des millions de petites boules creuses de cristal, dans la cavité desquelles on découvrit avec d'excellents microscopes, de petits corps qui se mouvaient continuellement autour d'un centre lumineux, comme les planètes se meuvent au tour du soleil, des espèces d'atomes dont les mouvements imitassent ceux des planètes. Ces petits globules paraîtraient d'abord d'admirable machines: ce serait une recherche digne d'un physicien de connaître le temps des révolutions de ces grains d'une prodigieuse petitesse, qui y seraient ce que sont les planètes dans le grand tourbillon, etc. . . ."**

* "Millions of small hollow crystal balls, inside of which would be discovered, through excellent microscopes, small bodies moving ceaselessly around a luminous center, just as planets move around the sun, that is to say, a kind of atoms, the motions of which would imitate that of planets. These small globules would, at first, strike us as admirable machines; it would be a worthwhile research for a physicist to find out the time of the revolutions of these grains so prodigiously small, acting there as the planets do in the great vortex, etc. . . ."

As we see, science in the seventeenth and at the beginning of the eighteenth century was above all Cartesian. But, little by little, it came under a new inspiration, this one brought from England; its mastery prevailed with an irresistible evidence, and its actuality is far from outgrown; this is the inspiration which emanated from Isaac Newton's works. The Cartesian phase was followed by the Newtonian. However, almost half a century was needed for Newton's conceptions to be well understood and accepted. Maupertuis was the first, in 1732, to support universal gravitation. Men like Réaumur and Fontenelle (1657-1757) remained Cartesian to the time of their death in 1757.

The publication, in Latin, of the *Principes mathématiques de la philosophie naturelle,* in 1687, marks one of the most capital dates in the advance of the human mind. But the importance of this great work, probably on account of its striking novelty, was but slowly recognized. In it, Newton formulates the principle of universal gravitation and establishes the fundamental equations of dynamics. Through these, he completes the work to which Galileo, Descartes and Huygens had made successive contributions. The conception of universal attraction was an even more striking novelty, which did not cease to dominate science. Let us remember, besides, that Newton shares with Leibnitz the glory of having invented infinitesimal calculus (the basis of which, it is true, had been laid down by Fermat and Pascal), from now on the essential and indispensable instrument of all exact science; if one adds also to Newton's credit his capital discoveries in optics (the decomposition and synthesis of white light, the explanation of the phenomena of interferences and the theory of light conceived as an emission of particles, a theory which ruled for a whole century), then one may reckon the unique place belonging to Newton in modern science.

That he never failed to receive justice in France is testified by the fact that, in 1699, the *Académie des Sciences,* when it reorganized, named him its first foreign associate member. Still, it took quite a long time for Newton's ideas to prevail in France and elsewhere. This should be no cause for great surprise. Attraction making itself felt at a distance, without any action of an intermediary medium, was, of course, bound to appear like a sort of occult force, recalling the scholastic phantoms, exorcised by Descartes; even to-day, it remains for us somewhat of an enigma, or at least, it is outside the pale of all the other forces of the physical world. Its reality could be impressed in no other way than through repeated probing of the consequences it led to; it is, mostly, in the field of celestial mechanics that this verification resulted in its triumph.

Now, it was in France, if not at once, at least in the end, that this verification was mostly made and brought about the most fertile discoveries. One of the first proofs of the Newtonian conceptions was the actual determination that the earth flattens at the poles as shown by the already mentioned expeditions to Lapland and Peru. Once accepted, Newton's ideas were more whole-heartedly embraced in France than anywhere else. Their acceptance extended far beyond the limits of the scientific world. Maupertuis was one of the first and enthusiastic exponents of Newton's ideas. Voltaire helped much to further their knowledge, notably through his *Eléments de la Philosophie de Newton mis à la portée de tout le monde,* and it was a woman, a mathematician of merit, by the way, the Marquise du Chatelet (1706-1749), who made a translation into French of the *Mathematical Principles of Natural Philosophy,* published after her death in 1756.

It is also in France, more than anywhere else, that

the consequences of the law of universal gravitation in the field of celestial mechanics have been developed. Laplace wrote:

"On doit à la France la justice d'observer que, si l'Angleterre a eu l'advantage de donner naissance à la découverte de la pesanteur universelle, c'est principalement aux géomètres français et aux prix décernés par l'Académie des Sciences, que sont dûs les nombreux dévelopements de cette découverte et la révolution qu'elle a produite dans l'astronomie, devenue la solution d'un grand problème de mécanique."[1]

Two great names personify in France the beginning of this great scientific movement, those of Clairaut (1713-1765) and d'Alembert (1717-1783), who were, with Euler, the greatest mathematicians of the middle of the eighteenth century.

Clairaut.

Clairaut had shown an astonishing precocity. At the age of twelve, he had read to the *Académie* his first memoir in which he already made use of infinitesimal calculus. Born on May 7, 1713, he was admitted to the *Académie des Sciences* as "adjoint mécanicien" on July 14, 1731 (he was 18 years old), "associé" on April 17, 1733 (at the age of 20), "pensionnaire mécanicien" on May 14, 1738. In the course of time, he made capital contributions to integral calculus and to the study of differential equations. Taking as a basis the measurements relative to the flattening of the earth, he published in 1743 his *Théorie*

---

[1] Laplace wrote: "One ought to be fair to France in noting that if England had the privilege of giving birth to the discovery of universal gravitation, French geometricians and prizes granted by the *Académie des Sciences* are mostly responsible for the numerous developments of this discovery and the revolution it caused in astronomy now considered as the solution of a great problem in mechanics."

de la Figure de la Terre. His mathematical researches on the equilibrium of a spheroidal fluid mass have remained the basis of all ulterior studies on the same subject. The same may be said of his work on the problem of three bodies, that is to say, the study of the motion of a heavenly body attracted by two others, a problem arising from the motions of the Earth, the Moon, and the Sun, or relating to the perturbations of planets as, for instance, Jupiter and Saturn. These researches, like those of d'Alembert and Euler, led to the perfect verification of Newton's law.

d'Alembert.[1]

d'Alembert's contribution was not less important than Clairaut's in the fields of algebra, infinitesimal calculus and mechanics. His *Traité de Dynamique,* published in 1743, is a piece of work of the first order. The principle of d'Alembert, which reduces the problem of a body in motion to the study of a body at rest, remains one of the fundamental conceptions of dynamics and provides a general method for equating, if not solving, all problems in that field. With the help of this principle, he successfully built up a whole series of fundamental works on the principal problems of mechanics.[2] He was also one of the found-

---

[1] A foundling, discovered on the steps of the church St. Jean le Rond (hence his name, Jean le Rond d'Alembert), brought up by a poor woman, he was admitted to the *Académie des Sciences* in 1741 (at 24) as "adjoint-astronome," and in 1754, elected to the French Academy of which he became "secrétaire perpétuel" in 1772.

[2] Here are the titles of some of these works: *Traité de Dynamique,* 1748; *Traité de l'équilibre et du mouvement des fluides,* 1744; *Réflexions sur la cause générale des vents,* 1746; *Essai d'une nouvelle théorie de la résistance des fluides,* 1746; *Recherches sur la précession des équinoxes et sur la nutation de l'axe de la Terre dans le système newtonien,* 1749; *Recherches sur différents points importants du système du Monde,* 1754-1756; *Opuscules mathématiques,* 1761-1780, 8 vol.; d'Alembert also was, with Diderot, the promoter and organizer of the *Dictionnaire Encyclopédique.*

ers of the mathematical theory of the efflux and flow of fluids, as well as of hydrodynamics. Not less considerable is his contribution to the study of the motion of the earth and the problem of three bodies. His researches in these fields are contemporary of Clairaut's but were independent of them as well as of Euler's.

By the side of Clairaut and d'Alembert, it is fitting to place Maupertuis, of whom we have already spoken, and to whom belongs the credit for the principle of lesser action, which he may have considered too much from the point of view of finality, but which Lagrange was to relate, as a consequence, to the very equations of motion.

*
* *

Physics in the Eighteenth Century.

Not less than mechanics, physics in the eighteenth century was Newtonian. But, in fact, it did not boast of the same fertile activity. Acoustics being reduced to the study of vibratile motions, benefited by the progress of infinitesimal calculus. In the history of physics we again find, following in Newton's footsteps, the same great names we have just mentioned in mechanics, notably d'Alembert. Rather strange to say, optics in the eighteenth century did not rise much beyond the point where Newton had carried it; there was never any thought that one could go further. Thus Bouguer (1698-1758) expressed it:

"Il peut sembler," dit-il, "qu'après toutes ces spéculations, il ne reste plus rien à découvrir dans cette matière. Cependant il est facile de s'apercevoir que, comme on s'est toujours borné jusqu'ici à examiner la seule situation et direction des rayons, il doit manquer encore à l'optique une partie tout entière qui aurait la force et la vivacité de la lumière pour objet."[1] This part was really photometry, which Bouguer fully constituted and exposed in his *Traité d'optique sur la graduation de la lumière* published after his death in 1760. In this direction only recently have we gone further than he did.

The eighteenth century saw the birth of electricity, under the form of electrostatics, which since has not led to important results, but which lent itself to striking experiments. The public in France took a great interest in this new science. We owe to Cisternay du Fay (1698-1739), superintendent of the *Jardin du Roi,* the discovery of the existence of two kinds of electricity, positive and negative. This was also the time of Benjamin Franklin's experiment which showed the way to the explanation of lightning and the invention of the lightning rod. Franklin's experiments were widely heralded in France and caused at once remarkable works along parallel lines in 1752; those of Dalibard and Delor at the instigation of Buffon and especially those of Jacques de Romas, who conducted experiments at Nérac, where he was a magistrate. Romas, while flying a kite at a great height and in the direction of storm clouds, drew on the rope sparks as much as six meters long; by directing the sparks to various bodies with the help of a discharger, he obtained impressive mechanical or thermal effects. He was a forerunner of Faraday in showing, at the same time, that electricity does not pene-

---

[1] "It may seem," said he, "that after all these speculations, there remains nothing to be discovered in this subject. Still it is easy to realize that, as investigation has always been limited up to now exclusively to the position and direction of rays, optics probably lacks a whole part dealing, as its subject, with the force and vivacity of light."

trate into a metal cage connected to the earth. An animal placed inside remains unhurt.

*

* *

French Society of the Eighteenth Century in Its Relation to Science.

As I bring this lecture to a close at a point near the middle of the eighteenth century, it is not without interest to leave aside for a while the exclusive study of great men of science and glance at the environment, institutions, and society in which they lived. French society in the eighteenth century was one of the most brilliant, polished and cultured the world has seen.

As the center of the scientific world and giving it its impulse stood the *Académie des Sciences,* the authority and brilliancy of which did not abate through the century. Its *Mémoires* and its *Histoire* regularly published each year, fill 114 volumes of the richest content concerning all sciences. Through its correspondents, each one keeping in touch with one of the members, the *Académie* was kept informed on all discoveries made abroad, even in the remotest countries beyond the seas. For instance, Abraham Trembley, observing at The Hague, in 1740, the multiplication by buds of the fresh-water hydra in the manner of a plant, Réaumur received from Trembley these particular animals and showed them to the *Académie;* so he wrote to Trembley:

"Je vous ai déjà marqué," écrit-il, "à Trembley, le 25 mars, 1741, que je me proposais de communiquer à l'Académie les observations qu'ils vous ont fournies et j'étais certain qu'elle leur donnerait beaucoup d'applaudissements. Je lui en fis part dès la première assemblée après votre lettre écrite. On ne se contenta pas de l'idée générale que j'en donnai tout d'abord.

On souhaita que je lusse toutes vos observations en entier et ce fut aussi unanimement que l'on donna des éloges à la manière dont elles sont rapportées et à l'attention avec laquelle elles ont été faites. La sage retenue qui parait dans toutes plut beaucoup. . . . Si l'on ne parlait pas trop de guerre à présent à Paris," écrit un peu plus tard Réaumur, "on n'y parlerait que des Insectes qui, étant coupés en deux deviennent des animaux complets. . . . Je racontai hier au Roi ce qu'il y a sur tout cela de certain et de merveilleux et Sa Majesté parut extrêmement contente de l'apprendre."[1]

We have seen the *Académie* organize great scientific expeditions like those which undertook the measurements of the meridian. The *Académie* thus contributed greatly to the advance of science and also helped it through the prizes awarded. Beginning with 1721, the *Académie* proposed important subjects and stimulated in this way researches of the most eminent men of science, not only in France, but also in foreign countries. Among the laureates we find such names as Jean and Daniel Bernouilli, MacLaurin, Euler, Lagrange.

Frédéric the Great, King of Prussia, moved by a desire to increase the brilliancy of the Academy of

---

[1] "I have already informed you on March 25, 1741, that I intended to communicate to the *Académie* the observations you have imparted to me and I was certain they would applaud them heartily. I presented them at the meeting following the receipt of your letter. They were not content with the general idea I unfolded at first. They wished me to read all your observations in full, and praise was also unanimously bestowed on the manner you have set them forth and the attention with which you conducted them. The wise restraint you show in all pleased very much. . . . If the talk of war was not so uppermost in Paris just now," wrote Réaumur, a little later on, "one would talk of nothing but insects, which after being cut into two become again complete living things. . . . I told the king yesterday all the certain and wonderful things we know about this and His Majesty looked extremely pleased to hear it."

Berlin, founded in 1700, at the instigation of Leibnitz and on the model of the Academy of Paris, drew to his capital some of the most illustrious Parisian men of science and even intrusted them with the direction of his Academy. Maupertuis, made famous by the expedition to Lapland, a great supporter of Newton's ideas, was thus called, in 1744, to direct it. The by-laws were drawn up in French, and in French likewise, were written the memoirs. When Maupertuis left Berlin in 1756, Frédéric urged d'Alembert to come and take his place, but d'Alembert did not accept; he only undertook to advise the King of Prussia from a distance and did so until his death. Condorcet succeeded him in this office. This shows the prestige French science was enjoying at that time.

The best Parisian society did not show a less keen interest in the sciences than in letters and arts. It is significant to see a lady of high birth such as the Marquise du Chatelet, translating the Principles of Newton. Experiments, particularly in electricity, were assiduously attended by "le grand monde." A very select audience crowded the lectures on physics of Abbé Nollet (1700-1770), at the Collège of Navarre, and the construction of a large amphitheater became necessary on this account. The same eagerness was seen at Rouelle's (1703-1770) lectures on chemistry at the *Jardin du Roi*. Diderot put in writing the lectures of this course and they were circulated in Parisian society; from these did Lavoisier get his taste for chemistry. The *Natural History of Buffon* met with the most favorable reception on the part of the public at large. Botany was no less in favor; people accompanied Bernard de Jussieu on field trips and this vogue was but to increase later under the influence of J. J. Rousseau. The publication of the great *Dictionnaire encyclopédique* directed by d'Alembert and Diderot vulgarized and spread scientific knowledge.

The refined and exacting tastes of this society had brought applied arts to a high degree of perfection, and every one knows what masterpieces of furniture we have inherited from that period. The all-around skill thus developed on the part of artisans reacted most favorably on the construction of scientific instruments of measure which reached a remarkable degree of perfection. Was it not the time when one saw Louis XVI devote his leisures to art-smithery?

So, by the side of the great creative minds, who were transforming science, an atmosphere favorable to its advance pervaded all through French society. As early as the middle of the century, such favorable conditions prepared the even more brilliant rise which was to illustrate the end of the eighteenth and the beginning of the nineteenth century.

March 8, 1933.

Second Lecture

SECOND HALF OF THE EIGHTEENTH CENTURY

———

Scientific and military schools.—Mathematics: Lagrange and Laplace. Monge and Legendre.—
Astronomy: Cassini's map.—Lavoisier and the
revolution in chemistry.—Lavoisier, founder of
general physiology.—Natural sciences: Buffon.
The Jussieus.—Geology and mineralogy.—The
great applications of science in the eighteenth century: the invention of the steamboat; aërostats.—
The metric system.—The Revolution and science.

———

The whole seventeenth century had been ruled by the
thought of Descartes. We saw that the influence of
Newton gradually substituted itself for it, during the
first half of the eighteenth century and reigned unrivalled during the second. The advance of celestial
mechanics brought an ever more striking confirmation
of Newtonian ideas, in regard to universal gravitation; the indispensable key to the study of all problems in mechanics and physics, infinitesimal calculus,
born at least in part from Newton's genius, was being
perfected, enabling thus new advances. We saw what
an important share was contributed by French mathe-

maticians, such as Maupertuis, Clairaut, and d'Alembert.

Scientific and Military Schools.

As early as the middle of the eighteenth century, it was realized that the major condition of any scientific development lies in a strong mathematical training. It is indispensable in the field of pure science, and equally needed in all the applications related to mechanics, such as those of public engineering and the military art. This fact was perfectly understood in France, without delay. In the absence of scientifically equipped universities, the need was satisfied by special schools; admission being obtained through competitive examinations requiring a high standard of mathematical proficiency, such schools always secured a brilliant body of students. Such were the *École Militaire* at Paris, and especially the *École du Génie* at Mézières, created in 1750, and the *École des Ponts et Chaussées,* organized in 1747 by Trudaine. Professors and students of these different schools provided, together with distinguished officers and engineers, a score of illustrious men of science. From the *École du Génie* of Mézières came Monge, Coulomb, Borda, Lazare Carnot, Meusnier, du Buat, Malus, etc. . . . Those schools therefore contributed efficaciously to the lustre of French science, at that time. They were the forerunners and the prototype of the schools which the Convention, in turn, organized, during the Revolution, and which, in the nineteenth century have been the framework of scientific life in France, at least in the field of exact sciences. It is interesting and but fair to recognize that the Monarchy whose frivolity and mistakes alone have generally been remembered, had, however, been wise enough to realize early the part science was going to play, and to create institutions favorable to its advance. With varying degrees of fortune, the men of science of this period witnessed and took a part in the political upheavals of the Revolution, the Empire and some of them in those of the Restoration; more than one played a rôle, at times considerable, outside of his scientific work; their lives were much more agitated than that of scientists, in general.

Mathematics: Lagrange and Laplace.

In the field of mathematics, two men of genius dominated the period: Lagrange (1736-1813) and Laplace (1749-1827).

Lagrange was born in Turin; his forbears were partly French. From early youth, he made himself known, and at the age of twenty-three, his reputation was great enough to have him elected an associate by the Academy of Berlin. In 1766, on the advice of d'Alembert, Frederic II sent for him to direct the section of sciences in this academy. He lived in Berlin until 1787, when he settled in Paris. The *Académie des Sciences* had made him an associate member as early as 1772. Welcomed with the greatest honors, naturalized a Frenchman, he was made a count of the Empire and on his death, in 1813, his remains were taken to the "Panthéon." Lagrange is one of the greatest figures in the history of mathematics. His work is distinguished by a particular character of perfection. His studies on the theory of equations opened the way to Abel and Galois. He carried on the work of Fermat on the theory of numbers; his researches on analytical functions, on linear differential equations, on partial differential equations of the first order, etc. . . . renewed those chapters of analysis. Lagrange gave to the general equations of mechanics their classical form; his *Mécanique analytique* is, for the initiated, a masterpiece, which Hamilton called a scientific poem. We must be content here with recording once more the unanimous admiration of mathematicians for Lagrange.[1]

Laplace, the son of a modest cottager in Normandy, came early to Paris, and caught the eye of d'Alembert with a letter on the principles of mechanics; thanks to the latter's help, in 1768, at the age of nineteen, he

Laplace

was appointed a professor at the *École Militaire*. In 1773, when twenty-four years old, he entered the *Académie* as "adjoint mécanicien." Since then, his life was filled with a series of successes and honors be-

[1] The *Œuvres complètes* of Lagrange have been published through the care of the *Académie des Sciences* (Paris, Gauthier-Villars, 14 volumes in 4°, 1867-1892).

stowed upon him by all the governments which followed each other during his lifetime. For a short while Bonaparte, then First Consul, named him *"Ministre de l'Intérieur."* Under the Empire, he was created a count, a marquis under the Restoration. Not only was Laplace a mathematician of the first rank, but his intelligence was of the most comprehensive type and remained vigorous even in old age.

Possessing a strongly synthetic turn of mind, he condensed the results of his original memoirs in general works the fame of which still survives. Such are his *Traité de Mécanique céleste, Traité analytique des Probabilités, Exposition du système du monde.* He really was one of those supremely intelligent men who dominate an epoch. It is, of course, quite impossible for me to attempt an examination of his mathematical work. It is enough to say that he completed in the eighteenth century the structure of celestial mechanics. He verified in detail and with the utmost precision, the perfect accuracy of Newton's law and applied it to the various celestial and terrestrial phenomena (such as tides). He developed the consequences of this law, both in their detailed and general aspects, and built them into a complete cosmogony, exclusively founded on mathematical analysis. The work begun by Newton has found in that of Laplace an extension and confirmation of limitless generality. Even this was not enough for his genius. We shall see him later, associated with Lavoisier in researches in physics, which were of capital importance for the study of heat; towards the end of his life he again gave to the new problems of the day a remarkable impetus. He, therefore, never stopped exerting all around a deep and beneficial influence.

This influence, in particular, made itself felt through those reunions he held in his home at Arcueil, a village at the gates of Paris; in 1806, he had bought there a piece of property, close by that of the chemist, Berthollet. We shall meet his name more than once later on.[1]

French mathematics, at the end of the eighteenth century, are far from being summed up in the works of Lagrange and Laplace. Among the other geometricians of this period, I shall limit myself to Monge and Legendre.

Monge and Legendre.

Gaspard Monge (1746-1818), the creator of descriptive geometry, was indeed a very beautiful personality. The son of an itinerant merchant, a brilliant pupil of the Fathers of the Oratory, *"puer aureus,"* he entered the *École de Mézières,* but his too modest birth forbade his becoming an officer and confined him in subaltern positions. The value of his works finally broke down the barrier of the social prejudice of his time. With the help of new methods, kept a secret until 1795, he renovated the art of fortification, and rendered invaluable services in the applications of mathematics to architecture and machinery, especially in the military field, during the Revolution. In 1795, he was one of the first and most eminent professors of the *École Polytechnique* and was deeply beloved by his students. A close friend of Bonaparte, whom he had helped when the former was a young officer, he accompanied him in the expedition to Egypt and became president of the *Institut d'Égypte.* The Emperor made him Count of Peluse. It is to be recalled to his honor that he remained obstinately faithful to Napoleon, a fact which accounted for his being one of the victims of the spite of the Restoration. He was crossed off the lists of the *Institut.* His

---

[1] The *Œuvres complètes* of Laplace have been assembled under the care of the *Académie des Sciences,* 6 volumes, in 4°, 1878-1892.

last years were filled with hardships. Descriptive geometry, which to this day stands as he constituted it, is the best known of his achievements, but it is far from being the only one; among other things, he brought remarkable advances in the geometry of surfaces studied through infinitesimal analysis.

Legendre (1752-1833), contrary to Laplace and Monge, lived through the troubled years of the Revolution and the Empire, without ever being turned aside from his daily scientific task. He, too, was a professor at the *École Militaire,* before the Revolution; he was admitted to the *Académie des Sciences* in 1783, and left an important mathematical contribution, inferior, however, to that of Lagrange, Laplace and Monge. It relates mostly to the theory of numbers and the theory of elliptical functions, a field in which his researches prepared the way for those of Abel and Jacobi.

*
*   *

Astronomy and Cassini's Map.

Observational astronomy, in the period we are dealing with, numbered several eminent representatives. I shall name, for instance: the Abbé de La Caille, (1713-1762), an able observer and calculator, who was one of the pioneers in the study of the austral sky, where he determined the position of over ten thousand stars; Méchain, (1744-1804), Lalande, (1732-1807), and Delambre, (1749-1822).

All through the century, the Cassinis ruled over the observatory, linking their name, outside of astronomy, to an undertaking which required long efforts, the map of France in 180 sheets, called Cassini's map. The enterprise had its beginnings in the seventeenth century, when the astronomer Picard proposed to Colbert, in view of drawing such a map, the plan of a general triangulation of France. The work was begun in 1683. The Cassinis kept it going on until the Revolution, through various interruptions and numerous financial difficulties. It is worth while to note, for instance, that in 1756, on account of the burdens of the war, the Royal Treasury had to stop supporting the undertaking. Cassini de Thury, (Cassini III), succeeded nevertheless, in continuing it, forming to do so, a company, which numbered among its forty stockholders: Buffon, the Duke of Luxembourg, Marshall of Noailles, Madame de Pompadour; this is one more significant testimony of the interest higher society took in the things of the mind. Finally, the work was accomplished. The map of Cassini was the first map of such a magnitude. It served as a model for all the great maps drawn at the beginning of the nineteenth century in several countries and its completion has helped a good deal the advance of geodetic science.

*
* *

Lavoisier and the Revolution in Chemistry.

Great as may have been the glory of mathematicians like Lagrange and Laplace, the glory which Lavoisier contributed to eighteenth century French science was at least no smaller. For he brought about a complete revolution and founded modern chemistry, that is, a science of unequaled fecundity in a field where mere empiricism and medieval chaos still ruled alone. In laying down the unassailable basic foundations of chemistry, Lavoisier took his place by the side of the greatest scientific geniuses whom he equaled.

With Lavoisier, (1743-1794), again we meet a man who was not a true professional scientist, though science was the chief interest of his life, and claimed the major part of his activity, since his youth. But, he also applied his tireless energy to numerous works of large scope, through much varied fields, and in fervent devotion to the common weal. He was just as much a great moral character as a great man of science. As the owner of an ample fortune, he had bought a share in the farming of the taxes and performed his duties in a scrupulous and methodical way. To his independent intellectual training he owed his lack of subservience to the commonly accepted ideas of his time. The soundness, clarity and depth of his judgment could have a free rein. It frequently happens, by the way, that men of this type, outsiders in fact, bring about great scientific revolutions.

As a young man, Lavoisier had taken a lively interest in very diversified branches of knowledge, notably in geology; in this field, some of his works bear the mark of his superiority and even go beyond their time. He also had touched upon zoölogy and astronomy. His first memoir was composed, in 1765, for a competition which the *Académie des Sciences* organized for the city of Paris, in view of improving lighting conditions. Lavoisier won a gold medal. But, from then onwards, he turned to chemistry for which he had acquired a taste from Rouelle's lectures at the *Jardin du Roi*.

Indeed, there did exist a sort of chemistry before Lavoisier, but it was entirely empirical and in regard to general conceptions still abiding by the alchemy of the Middle Ages, as the language in use gives proof. The four elements still served as its basis: earth, water, air, and fire. It was alive with those phantoms which Descartes had exorcised, and full of verbal principles which clouded phenomena. For this, Lavoisier substituted a body of notions founded on strict experience and the evidence of facts. The point on which he attacked the old conception was the question of transformation of metals into *earths* or *calx,* to use the term then current, a phenomenon, we have come to know since Lavoisier, as oxidation. It was then explained according to a theory devised at the end of the seventeenth century by a German physician, Stahl. This theory held absolute sway, and was accepted as positive truth, giving the key to everything: it was the theory of *phlogiston.*

Any kind of body when burnt released an immaterial and universal principle: *phlogiston;* likewise, in calcining, in the open air, a metal like lead or tin, one deprived it of its *phlogiston.* The *calx,* thus formed, was the metal *dephlogisticated.* Conversely, by heating *calces* with a body rich in *phlogiston,* like charcoal (this, for us, means reducing the oxide) *phlogiston* was given back to *calx,* and the metal was thus reconstituted.

Lavoisier

It would be extremely interesting to relate in detail the series of steps by which Lavoisier overthrew that whole theory which lacked a real basis; he did so through experiments as simple as they were ingenious, conducting them with impeccable logic, and utilizing also experiments which chemists who believed in *phlogiston*—some of them were eminent men—had made in quite a different spirit. We have here one of the most fascinating chapters in the history of science and the emancipation of the human mind. In carrying out this magnificent work, Lavoisier had but applied the rules of the Cartesian method, proceeding strictly from the known to the unknown, and only trusting well established facts. But I cannot, in the time at my disposal, enter into such an examination, and I refer those whom this account might tempt to the admirable book, which in our days, Ed. Grimaux has devoted to Lavoisier's work.[1]

I shall merely say that the fundamental and irrefutable fact was the demonstration that metals, when transformed into *calx,* far from losing anything, on the contrary, increased in weight,[2] the increase being caused by their absorbing from the air a substance, without which the air became unfit for breathing. This substance, Lavoisier called it, at first, the "most respirable air" *(l'air éminemment respirable)* and later oxygen—Lavoisier was fully conscious of the importance of the result. As early as 1773, he himself declared that it amounted to a revolution in chemistry. He was then thirty years old. For immaterial principles, he substituted real, tangible bodies which he detected and measured with the help of a precision instrument, the balance. From then onwards, the balance became he fundamental tool of the chemist.

Lavoisier's capital discoveries multiplied rapidly in the course of a few years. Of these, let us recall only that air, far from being an element, was recognized as a mixture; that water, too, ceased being an element and was revealed, in fact, as a compound. Thus, in

[1] Ed. Grimaux, *Lavoisier* (1743-1794) *d'après sa correspondance, ses manuscrits, ses papiers de famille et d'autres documents.* Paris —1888.—The *Œuvres complètes de Lavoisier* have been collected in 6 volumes in 4°, *Imprimerie Nationale,* 1862-1893.

[2] This very important fact concerning the increase in weight of metals by calcination had been clearly observed as early as 1630, by a physician from Périgord, Jean Rey, *(Essais sur la recherche de la cause pour laquelle l'estain et le plomb augmentent de poids quand on les calcine,* Bazas, 1630). This memoir was completely forgotten. It has been re-edited by Ed. Grimaux (Paris, 1896, Masson). Jean Rey was certainly a forerunner of genius.

about ten years, an entirely new chemistry was constituted, the basis of which was, first of all, a knowledge of the nature of gases and air and which, for this reason, was called *pneumatic chemistry*.

What shows better than anything else, the fundamental importance of Lavoisier's work, which even to our day German chemists, like Ostwald, have insidiously tried to depreciate, what proves the novelty of his discoveries, was the violent, unanimous and continued opposition it met at the hands of all chemists, both French and foreign, including the most eminent: Scheele in Sweden, Priestley, Cavendish, and Kirwan in England, Macquer, Fourcroy, Berthollet, Guyton de Morveau, in France. From 1775 to 1785, as Fourcroy wrote later, Lavoisier was alone of his opinion, in spite of the clarity of his memoirs and the illuminating logic of his experiments. Things turned out again just like in the case of the opposition to the discovery of the circulation of the blood, by Harvey, in the seventeenth century. The same humorous note was not even missing. We listened to the imprecations and wailings of Riolan against Pecquet. In the same fashion, Macquer, a chemist of renown and a man of good faith, wrote to Guyton de Morveau, in 1778: "M. Lavoisier m'effrayait depuis longtemps par une grande découverte qu'il réservait, *in petto,* et qui n'allait pas moins qu' à renverser de fond en comble toute la théorie du phlogistique, ou feu combiné. Son air de confiance me faisait mourir de peur. Où en aurions-nous été s'il avait fallu avec *notre vieille chimie,* rebâtir un édifice différent?"[1] To refute the arguments of Lavoisier and maintain *phlogiston,* Guyton

---

[1] "M. Lavoisier had long been frightening me with a great discovery which he reserved *in petto,* and which aimed at no less than the complete overthrow of the whole theory of *phlogiston* or combined fire. His air of confidence had me dying from fear. Where would we have been if we had been obliged with *our old chemistry,* to build up a different structure?"—Cf. p. 21.

de Morveau did not hesitate to admit that it must have a negative weight, thus explaining, he said, the increase in weight of the metal through calcination. The nineteenth century, in turn, will hear discussions of this brand, out of the mouths of physicians, this time, while the pastorian revolution was being enacted.

When, going further, Lavoisier, in 1783, had demonstrated that water is not a simple element, this great and illuminating discovery, which was established both by analysis and synthesis, brought him nothing but a renewal of attacks from the upholders of *phlogiston,* including the most celebrated, like Priestley.

Not before 1785, did he make his first convert, Berthollet, who was soon followed by Fourcroy, and later by Guyton de Morveau; the three of them joined Lavoisier in establishing, on the basis of the new ideas and facts, the chemical nomenclature still in use to-day. In 1789, Lavoisier published a *Traité de Chimie,* an elementary textbook, which then was received with huge applause and marked the triumph of the new science as well as the final wreck of Stahl's chemistry. The novelty of the language used in the book is, in itself, an index to the revolution which had taken place. From that point, the greater part of the chemists assented to the doctrine, particularly Black, who, even before Lavoisier, had had the very great merit of making decisive advances in the handling of gases and had isolated *fixed air,* our carbon dioxide. But there remained some die-hards and among the most notorious were Priestley, Cavendish, and Scheele. The revolution brought about by Lavoisier went far beyond the limits of the chemistry of his day. He formulated a new and rational conception of matter and its transformations; such a conception reflected on the whole universe, and the *principle of the conservation of matter* represented an advance comparable to that of universal gravitation; Lavoisier expressed it thus in his *Traité de Chimie,* in 1789, in reference to alcoholic fermentation: "Rien ne se crée, ni dans les opérations de l'art, ni dans celles de la nature et l'on peut poser en principe que, dans toute opération, il y a une égale quantité de matière avant et après l'opération. C'est sur ce principe qu'est fondé l'art de faire des expériences en chimie. On est obligé de supposer dans toutes une véritable égalité ou équation entre les principes des corps qu'on examine et ceux qu'on en retire après l'analyse."[1]   A familiar and self-evident truth nowadays, but which was then a capital novelty.

Lavoisier, Founder of General Physiology.

The work and influence of Lavoisier were not limited to chemistry. They left their mark almost as deeply in physics and physiology. Ever since his first researches, his attention had been drawn to the fact that that part of the air necessary to respiration is at the same time the agent of combustion. At first, he called it *l'air éminemment respirable,* (respirable air), later *oxygen.* In a memoir of 1777, *Sur la respiration des animaux et les changements qui arrivent à l'air en passant par les poumons,* he established the capital fact, just as important in physiology as his novel ideas were in chemistry, that respiration is a kind of combustion, a slow combustion taking place inside the body, in the lungs or in some other organs; he did not decide this question of localization. The result of respiration, as well as that of any combustion, is the production of *fixed air* (carbon dioxide), that

---

[1] "Nothing creates itself, neither in the operations of art, nor in those of Nature and one may lay down as a principle that in any operation, there is an equal quantity of matter before and after the operation. It is on this principle that the art of making experiments in chemistry is founded. One is compelled to assume in all of them a real equality, or equation, between the principles of the bodies under examination and those derived from them, after the analysis."

same gas which Black had previously obtained and isolated from the calcination of limestone.[1]

Now, therefore, one of the most fundamental operations of life, respiration, was henceforth understood; its correct interpretation displaced the age-old and erroneous opinion of Aristotle, still current, according to which respiration refreshed blood. At the same time, an essential vital phenomenon was reduced to terms of physico-chemical mechanism. But this was merely a starting point. Lavoisier perceived immediately in respiratory combustions the source of animal heat and applied himself to the study of this phenomenon in a precise way, through quantitative measurements. In this investigation, he coöperated with Laplace; their memoir, published in 1780, established the technique of the measurement of quantities of heat, and thus founded a new and important chapter in physics, calorimetry.

In this memoir, written in collaboration with Laplace and in other memoirs written in 1785 and 1788 with Seguin, Lavoisier discovered the caloric part played by transpiration, which is the regulating function of animal heat. He concludes thus: "La digestion rend à la machine tout ce qu'elle perd par la transpiration et la respiration."[2] Energetic biology, as a whole, was thus constituted, at a single stroke, a science into which the nineteenth century was to delve more deeply. The terms used by Lavoisier in the following lines show how clearly he brought out these fundamental truths: "C'est l'air de l'atmosphère," écrit-il, en 1789, "qui

---

[1] This wonderful discovery of Black had not met with unanimous assent in its interpretation. A pharmacist from Osnabrück, Meyer, had received great applause for attributing the causticity of calcined *calx* to a special principle, *acidum pingue,* a phantom of the same vintage as *phlogiston.*

[2] "Digestion gives back to the machine all that one loses through perspiration and respiration."

fournit l'oxygène et le calorique, mais c'est la substance de l'animal, c'est à dire le sang, qui fournit le combustible et, si les animaux ne réparaient pas habituellement par les aliments ce qu'ils perdent par la respiration, l'huile manquerait bientôt à la lampe et l'animal périrait comme une lampe s'éteint quand elle manque de nourriture. . . . En rapprochant ces résultats de ceux qui les ont précédés, on voit que la machine animale est principalement gouvernée par trois régulateurs principaux : la respiration qui consomme de l'oxygène et du carbone et qui fournit le calorique, la transpiration qui augmente ou qui diminue, suivant qu'il est nécessaire d'emporter plus ou moins de calorique, enfin la digestion qui rend au sang ce qu'il perd par la respiration et la transpiration."[1]   All this was established, not on speculations and hypotheses, but resulted from irrefutable facts, and no part of it has since then been shaken down.

Lavoisier perceived even vaster horizons, as may be ascertained from the argument of a prize essay which the *Académie* proposed in 1789, for the year 1794. The argument is anonymous, but the draft was found in Lavoisier's papers : "Les végétaux," dit-il, "puisent dans l'atmosphère qui les environnent, dans l'eau et en général, dans le règne minéral les matériaux nécessaires à leur organisation.   Les animaux se nourrissent, ou de végétaux ou d'autres animaux, qui ont été

---

[1] "It is the air of the atmosphere," he wrote in 1789, "which furnishes oxygen and caloric, but it is the substance of the animal, that is to say, blood which furnishes the combustion, and if animals did not habitually regain with food what they lose through respiration, the lamp would soon lack oil and the animal would perish just like a lamp goes out when short of fuel. . . . In comparing these results with the preceding ones, one sees that the animal machine is principally governed by three principal regulators : respiration, which consumes oxygen and carbon and furnishes caloric ; perspiration, which increases or diminishes, in proportion as it is necessary to take away more or less caloric ; finally, digestion, which gives back to the blood that which it loses through respiration and perspiration."

eux-mêmes nourris de végétaux, en sorte que les matériaux qui les forment sont toujours, en dernier résultat, tirés de l'air et du règne minéral. Enfin la fermentation, la putréfaction et la combustion rendent perpétuellement à l'air de l'atmosphère et au règne minéral les principes que les végétaux et les animaux en ont empruntés.

"Par quels procédés la Nature opère-t-elle cette merveilleuse circulation entre les trois règnes? Comment parvient-elle à former des substances combustibles, fermentescibles et putrescibles avec des matériaux qui n'avaient aucune de ces propriétés? Ce sont jusqu'ici des mystères impénétrables. On entrevoit cependant que, puisque la combustion et la putréfaction sont les moyens que la Nature emploie pour rendre au règne minéral les matériaux qu'elle en a tirés pour former les animaux et les végétaux, la végétation et l'animalisation doivent être des phénomènes inverses de la combustion et de la putréfaction."[1]

Views of an astonishing breadth and prophetic depth on the phenomena of life! It remained for Pasteur three quarters of a century later, to meditate upon them and lift the veil of mystery.

---

[1] "Plants," he says, "draw from the surrounding atmosphere, from water and, generally speaking, from the mineral kingdom, materials necessary to their organization. Animals feed either on plants or other animals, themselves fed on plants, so that materials forming them are always in final analysis drawn from the air or from minerals. Finally, fermentation, putrefaction and combustion perpetually give back to the atmosphere and the mineral kingdom, the principles borrowed from them by plants and animals. By what means does Nature work out this marvelous circulation between the three kingdoms? How does she succeed in forming substances combustible, fermentable and putrefiable with materials formerly deprived of all these properties? Those questions are still impenetrable mysteries. One gets a glimpse, however, that since combustion and putrefaction are the means Nature uses to give back to the mineral kingdom materials drawn from it to form animals and plants, conversely vegetation and animalization must be inverse phenomena to combustion and putrefaction."

The Revolution and Lavoisier's Death.

All this work of Lavoisier was achieved from 1773 to 1789, between the ages of thirty to forty-six. After the period of struggles, he had reaped all the glory he deserved; his authority was acknowledged by all his contemporaries. His town house at L'Arsenal, his laboratory were the meeting places of French and foreign chemists and mathematicians. Under normal circumstances, he, no doubt, would have brought more capital contributions to science and hastened its progress. But, in 1789, the Revolution broke out. The train of events tore off Lavoisier from his laboratory. He was called upon to study various important questions; he devoted himself to the common good and rendered immense services. Came the Terror. The *fermiers généraux,* tax collectors of the Monarchy, bore the cost of the fiscal abuses of the old régime; they were arrested, brought before the Revolutionary Tribunal, and after a parody of justice, sentenced and put to death within twenty-four hours. Lavoisier was engulfed into the catastrophe and guillotined with twenty-six others, among whom was his father-in-law, Paulze, on May 8, 1794. One must acknowledge with sorrow that none of the voices which should have been raised to defend him and keep him safe for science and for his country made itself heard. "Son crédit," a écrit, l'astronome Lalande, "sa réputation, sa fortune, sa place à la Trésorerie, lui donnèrent une prépondérance dont il ne se servit que pour faire le bien, mais qui n'a pas laissé de lui faire des jaloux. J'aime à croire qu'ils n'ont pas contribué à le perdre."[1]

Silence on the part of men like Fourcroy, then a

---

[1] "His credit," wrote the astronomer, Lalande, "his reputation, his fortune, his position at the Treasury, gave him an eminence which he never used for anything but to do good, but, nevertheless, made some people feel jealous. I would like to believe they did not help to bring about his downfall."

member of the Convention, is an horrible stigma on their memory. The death of Lavoisier is one of the most dreadful dramas in the history of sciences and one of the most indelible stains that sullied the Revolution.

*

* *

Natural Sciences.—Buffon.

The works of Lavoisier have carried us into the field of physiology and so lead us to glance at the natural sciences. I am obliged to limit myself to the two names most representative of the epoch: Buffon and the Jussieus.

Buffon (1707-1788) enjoyed during his lifetime a great celebrity; his *Histoire naturelle générale et particulière,* the thirty-six volumes of which were published from 1749 to 1788, was a powerful factor in spreading a taste for zoölogy and acquainting the public at large with the general problems of Nature. Buffon's early training was in mathematics and he was well acquainted with Newton's doctrines; chance was responsible for directing him toward natural sciences. In 1738, the superintendent of the *Jardin du Roi,* Cisternay du Fay, a physicist of merit, who, as I said before, discovered the distinction between the two kinds of electricity, prematurely died of smallpox; on his deathbed he recommended to the king that Buffon be made his successor. Buffon kept the direction of the *Jardin* for half a century. As he was not a naturalist, he strove to become one and considered it a duty of his office to study and describe Nature in its general features, to draw up a picture of the whole subject, encompassing at the same time, the main lines and the detail of facts, a picture whence could be deduced the philosophy of Nature. The style of the work is pompous and grandiloquent. Leaving to his assistants, Daubenton, the Abbé Bexon, Guéneau de Montbéliard, the care of collecting a mass of documents and specific facts, Buffon, without neglecting these details, devoted himself mostly to the study of the general problems. No striking discovery is connected with his name, but his undeniable merit was to perceive first, through a series of bright intuitions, a large number of ideas and conceptions which formed the basis of nineteenth century zoölogical philosophy. These ideas were then new and bold enough to worry theologians. The all-powerful *Sorbonne,* which was the Faculty of Theology of the University of Paris, broke forth against the *Théorie de la Terre,* which appeared in 1749, as the first volume of Buffon's *Histoire Naturelle.* Buffon, among other things, was guilty of estimating the age of the earth at 60,000 years instead of 6,000, as did Biblical tradition. The *Sorbonne* condemned fourteen propositions extracted from the book. Buffon, who did not care to engage in a fight with the Church, yielded, at least outwardly; however, his thought remained intrinsically and unmistakably committed to an explanation of the world through purely natural laws, and he may be considered a forerunner of transformism.

His long directorship of the *Jardin du Roi* imparted a great brilliance to this establishment and was particularly fruitful in furthering expeditions of scientific research. At that time, the tropics were much less easy of access than now; France, though, had already built up a diversified colonial empire including India, the Mascareigne Islands, Guiana, Canada, Louisiana, and French naturalists traveled extensively through the world. My colleague, M. Alfred Lacroix, has just presented this year to the *Académie des Sciences* an account of their work in the West Indies and Guiana. They did considerable work. Réaumur was receiving important shipments from his numerous correspondents beyond the seas. The *Jardin du Roi* had on its pay roll numerous missionaries, such as Poivre, Sonnerat, Commerson, Adanson, and many others. His Majesty's government organized many expeditions to the Pacific, like that of Bougainville, which brought back much valuable information about Oceania and revealed Tahiti to the world; that, of la Pérouse, in 1788, whose ship was completely lost on the reefs of Vanikoro. The *Jardin du Roi* enjoyed a great prestige during the period. Its collections were the richest in Europe. Buffon was treated with the utmost regard by monarchs like the Empress of Russia, Catherine the Great.

The Jussieus.

The part played by the *Jardin du Roi* in the field of botany was no less brilliant, owing to the Jussieus. Bernard de Jussieu, a modest man, was one of the best naturalists of his time and, especially, a peerless botanist. He remained, from choice, in the shadow of his brother, Antoine, and of his nephew, Antoine-Laurent, satisfied all his life with the rank of "demonstrateur," that is, of assistant; in reality, he was the inspiring force in botany at the *Jardin*. Through his teaching, he exercised a fertile influence. His profound knowledge of plants enabled him clearly to perceive and identify the affinities between the great natural families. It was according to such affinities, that breaking away from tradition, he arranged the collections of live plants, first, in the royal garden of *Trianon,* later, in Paris at the *Jardin du Roi.* Antoine-Laurent de Jussieu completed the work of his uncle and perpetuated it by publishing, in 1789, his celebrated *Genera plantarum secundum ordines naturales disposita.* The principle of the subordination of characters, the basis of this classification, exerted a considerable influence, not only on botany, but also on zoölogy, at the beginning of the nineteenth century. It is at the origin of the great conceptions of Geoffroy Saint-Hilaire and Cuvier.

Geology and Mineralogy.

To complete the review of the sciences, at this epoch, one must also say a few words on geology and mineralogy. The word geology, created by de Luc, from Geneva, precisely appears for the first time at the end of the eighteenth century. It is inscribed on the frontispiece of the *Encyclopédie* in the genealogical table of the sciences. Geology was not to begin its real career until the nineteenth century, but it had pioneers in France, notably in Guettard (1715-1786) and Nicolas Desmarets (1725-1815), and a score of paleontologists. Guettard had developed an interest in geology on the part of young Lavoisier, who, around his twentieth year, took with him an extended trip through eastern France. Lavoisier, by the way, did some work in geology, and what he published on this subject, notably on the Paris basin, bears the mark of his mastery.

It is, likewise, at the end of the eighteenth century, that mineralogy was constituted as a modern science through the advance of crystallography. From all time, minerals had attracted the curiosity of collectors and some minerals had already played an important scientific part, such as Iceland spar, on which, in the seventeenth century, Huygens had discovered and studied dual refraction. Though there had been precursors like Romé de l'Isle (1736-1795), the author of a mineralogy published in 1772, the creator of crystallography was the Abbé René Haüy, (1743-1822). He recognized in the geometrical forms of crystals and in the laws of their symmetry, the expression of essential properties of their elementary structure, of their constituting particles. Thus was expressed the problem of the molecular structure of crystallized matter, a problem which has played and still plays such an important part in physics. By the side of Haüy, it is fitting to recall the name of another mineralogist, the Chevalier Déodat de Dolomieu, (1750-1801), of the Knights of Malta, a companion of Bonaparte during the expedition to Egypt; on his return to France, the chevalier was taken a prisoner by the English and died shortly after his release, from hardships endured during twenty-one months of a painful captivity. Dolomieu has given his name to a well-known rock, dolomite.

*
* *

The Great Applications of Science in the Eighteenth Century.

In all that precedes, we have examined only pure sciences. But their advance was beginning to be translated into applications which, in the following century, were to transform so deeply the conditions of society. The *Académie des Sciences* did not neglect them, particularly those of mechanics; it collected and published drawings of machines. For a century there had been produced works on hydraulics, both theoretical and practical. Hydraulics counted eminent representatives

The Steam Carriage of Cugnot

like Bélidor, Chézy, du Buat. The first development of the steam engine dates back to that time; then also appeared the first and modest ancestor of the automobile, the steam carriage of Cugnot, built in 1769 to draw guns. Its speed was but four kilometers an hour and it could not run for more than fifteen minutes without stopping. An improved carriage was built in 1771, under the auspices of the Minister Choiseul, but he fell from power and his successor did not take any interest in the matter; the unused carriage still remains in the collections of the *Conservatoire des Arts et Métiers.*

The Invention of Steam Navigation.

It is also from that time that date the first practical attempts of steam navigation, the real inventor of which was the Marquis de Jouffroy d'Abbans. In 1776, a sailless and oarless boat, provided with a steam engine, plied for two months on the Doubs, near Besançon; another, 43 meters long, the "pyroscaphe," equipped with paddle wheels, plied for several months on the

The "Pyroscaphe" Navigating the Saône at Lyons

Saône, at Lyons, in 1783. A third one was built at Paris to be launched on the Seine, but lack of financial resources preventing its completion, it never was finished. A few years after Jouffroy d'Abbans's trials, the problem was taken up by an American, Fulton, who vainly tried to interest Bonaparte in the experiments he made on the Seine in 1799. He repeated them, successfully, later, in 1812, on the Hudson, at New York. They were the real starting point of steam navigation. But the priority of Jouffroy d'Abbans, to whom Fulton did not fail to pay homage, must not be forgotten.[1] Jouffroy d'Abbans met with the fate of many inventors. Ruined by the Revolution and the emigration, reduced to dire poverty, he died in 1832 at the *Invalides,* where he had been admitted in his old age.

Aërial Navigation.

The invention of balloons and the beginnings of aërial navigation were much more brilliant; the success was immediate and resounding.

The inventors were the Montgolfier brothers, paper manufacturers from Annonay, in Ardèche. On June 5, 1783, on the day of the assembly of the États of Vivarais, at Annonay, they inflated with hot air an envelope of sack cloth lined with paper and weighing 500 pounds. It rose to 1,000 "toises" (about 2,000 meters). This experiment was communicated to the *Académie* and brought forth enthusiasm. It was repeated in Paris, a few weeks later, by the physicist, Charles, (August 27, 1783), but this time the envelope was made of silk. The balloon, of a capacity of 40 cubic meters, was inflated with that *inflammable air*

---

[1] Fulton had another forerunner in New York, John Fitch, whose paddle-wheeled boat, propelled by steam, navigated on Collect Pond in 1793. Fitch's experiments, however, came sixteen years later than those of Jouffroy d'Abbans.

(hydrogen) Lavoisier had just discovered; it rose up in a few minutes to about 1,000 meters and there burst. On the 21st of November of the same year, the first ascension was carried out by Pilâtre de Rosier and the Marquis d'Arlandes, in a balloon inflated with hot air, according to the method of Montgolfier. This balloon, starting from the park of la Muette, sailed across Paris and landed at la Butte aux Cailles. Charles, who firmly believed in the use of hydrogen, in a few weeks, imagined all the equipment of modern balloons (envelope impermeable to gases, car, net, valve, barometer to measure the altitude). The quick securing of so novel an equipment testifies to the efficiency of the industry, technique and workmanship of the times. Finally, on December 1st, Charles and a passenger, Robert, rose up in this balloon from the garden of the Tuileries, before a huge crowd, and landed at l'Isle Adam, after two hours of aërial navigation. In six months, this invention had been perfected.

More remarkable still, the whole theory of aërostats and even that of dirigibles which were not to be built for another hundred years, was immediately framed. On December 3, 1783, a lieutenant of engineers, Meusnier (1754-1793), who unfortunately was to be killed in 1793, at the age of thirty-eight, during the siege of Mayence, read to the *Académie* his *memoir on the equilibrium of aërostatic machines, on the different ways to make them rise and come down and especially on executing these maneuvers without throwing ballast and losing inflammable air.* This memoir not only contains the technique of spherical balloons, but also the principle of contrivances which were not made available until our own days. Such was the use of small inside balloons filled with air, an idea to which in 1870, Dupuy de Lôme reverted, in his project of a dirigible. Meusnier conceived of an envelope with the shape of an ellipsoid and he thought of directing the ship by means of screw propellers actioned by the crew. The balloon he imagined was to have a crew of 24 sailors and 6 officers carrying food rations for

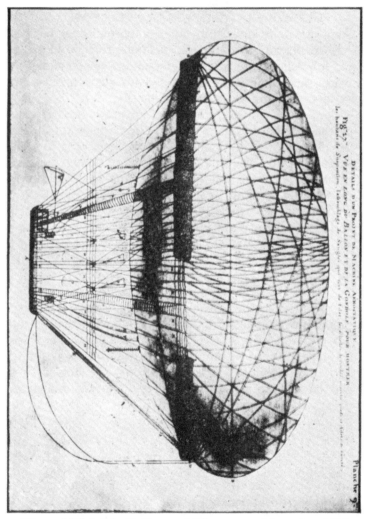

Plan of Dirigible Balloon of Lieutenant Meusnier, 1783

66 days. All in all, Meusnier, in 1783, drew up the whole theory of modern aëronautics. He was an eminent man of science, fresh from the *École de Mézières,* where as a student he had aroused Monge's admiration by completing a theory of Euler. At the age of twenty-two, he was a correspondent of the *Académie* and at twenty-nine, a regular member. It was to obtain hydrogen for balloons that he had been led to collaborate with Lavoisier in the analysis of water and the production of hydrogen through the action of steam on red-hot iron.

Meusnier was a remarkable inventor and a great mathematician. The events of the Revolution, unfortunately, turned him away from science and his premature death prevented him from giving his full measure. Monge considered him the most extraordinary mind he had ever known. Meusnier deserves a distinguished rank in the phalanx of great Frenchmen of science, at the end of the eighteenth century.

*
* *

The Metric System.

The metric system was another magnificent application of the advance of exact sciences during that period. The system did not become a reality until the extreme end of the century, though preparatory work had been carried on almost from its beginning. This reform was born from a need to unify the weights and measures in use in the several provinces. At the time of his death, in 1738, Cisternay du Fay, superintendent of the *Jardin du Roi,* was busy on this work. Studies were resumed in 1766 by order of the Minister Trudaine. In view of the adoption of the same basic unit everywhere in France and even outside of the country, efforts tended, from the start, to define the unit not arbitrarily, but in relation to a natural phenomenon. The astronomer Picard, in the seventeenth century, had at first thought of the length of the pendulum striking the second, but it was soon found out not to be the same in all places. One was led to think of a definite fraction of a quadrant of the earth; from this idea, the metre was born. All these projects show the craving of the French mind for logic and precision. On the other hand, thinking people were strongly impressed with the convenience of the decimal system. At the end of the eighteenth century, men of science used it in their calculations, to the exclusion of the duodecimal subdivisions of the common units. It is to be remarked, by the way, that nowadays, almost a century later, we are witnessing a similar practice in Anglo-Saxon countries, which have not yet been able to free themselves of empirical traditionalism and adopt the metric system; in English and American laboratories, the decimal system and metric units are almost everywhere in use for calculations, whereas the old ways still prevail in business and practical life.

On the eve of the Revolution, the reform of the system of weights and measures was requested in most of the memorials, *(cahiers)* drawn up for the State Generals; one reads in the report of the Marquis de Bonnay, a deputy for nobility: "L'Angleterre est prête à se joindre à nous pour exécuter cette uniformité. Quand ces deux nations qui n'ont de rivales qu'elles-mêmes l'auront adoptée, toute l'Europe ne manquera pas de l'adopter aussi."[1] Europe has fulfilled the prediction, though neither England nor the United States have.

The "Assemblée Constituante" on the report of the bishop of Autun, Talleyrand, very wisely entrusted the care of establishing the basis of a new system of units

---

[1] "England is ready to join us in establishing this uniformity. When these two nations who have no rivals but themselves have adopted it, Europe will not fail to adopt it also."

to the *Académie des Sciences*. A commission appointed by the *Académie*, selected as the fundamental unit of length, the metre equal to one ten millionth part of a quadrant of the earth, as the unit of mass, the gramme, the mass of a cubic centimeter of distilled water at 4° C. The commission outlined the work to be done to establish the new units. It worked with the greatest zeal during the months following its organization, which were the last of the *Académie*, suspended by order of the Convention, in August, 1793. Lavoisier was the inspiring force on the commission. To determine the length of the metre as accurately as possible, it was decided to proceed to a new measurement of the meridian, between Dunkirk and Barcelona, this time. All the foremost scientists took part in the work: Delambre, Lagrange, Laplace, Monge, Legendre, Cassini, Méchain, Meusnier, and two other scientists of the first rank whose names have not been mentioned yet, Borda and Coulomb, who were in charge of measurements of the pendulum.

The Chevalier de Borda (1733-1799), came from the *École du Génie de Mézières*. Since youth, he had distinguished himself by works on physics, hydraulics, and geodesy; circumstances led him to enter the navy, in which he had made a brilliant career. In particular, he had been major general of Admiral d'Estaing's fleet, which was sent to help the American insurgents in 1778. He left a deep mark in the organization of the French navy. Borda contributed much to the improvement of measuring instruments. It was with the help of a divided circle, built on his indications, that the angles of the new triangulation were measured. His devotion to science may be reckoned by the fact that at the end of his life, wishing to establish trigonometric tables in the decimal system, he had the calculations made at his own expense and sold one of his lands to insure the publication.

The work preparatory to the reform lasted from 1791 to 1798. When it was ended, the nations friendly to France were invited to send delegates to Paris in view of being acquainted with the results, of joining in their completion and in the final determination of the new units. The state of war with England was responsible for her not being represented at this reunion. The standards of the metre and the kilogram were presented, with solemnity, to the legislative bodies of the time, the *Conseil des Anciens* and the *Cinq Cents* by a delegation of the *Institut,* on the 4th messidor, an VII (June 22, 1799). These standards, henceforth, constituted the actual units. The metric system, however, was not made compulsory at once; it only became actually so in 1840. The value of this work, inspired by purely scientific consideration, has been proved in the last hundred years, by the extension of the metric system to almost the whole globe, with the exception of the Anglo-Saxon countries, and by the ease with which nations have become used to it. Since 1872, an *International Bureau of Standards (Bureau International des Poids et Mesures),* the seat of which is at Sèvres, near Paris, is entrusted with the preservation of the standard units and with the performance of all work related to them. Although the United States have not yet adopted the metric system, illustrious American men of science like the late lamented Michelson have contributed to metrical works of high precision in the laboratories of the Bureau.

The Revolution and Science.

The above outline has necessarily been restricted to trace the French scientific movement in the second half of the eighteenth century, only in its more general aspects; it has taken into account only men and works of the first rank. This has been sufficient, so it seems, to show its lustre and radiant diffusion. French civilization, at that time, was really ruling over Europe. The brilliant society, of which even the more frivolous elements showed a real understanding and appreciation for scientific matters, was thrown in a state of turmoil, by the Revolution, from 1789 onwards. This is not the place to examine the deep-seated and legitimate causes of the Revolution. The American War of Independence certainly helped to bring it about, in exalting the spirit of freedom; one is aware, for instance, of the great popularity Benjamin Franklin enjoyed in Paris. Nor do we need to sum up the political and social sides of the ledger. But we may and, in my opinion, we should briefly examine what effect the Revolution had on science. It is to be noted, first, that from the very beginning most of the greater scientific minds openly rallied to the side of the Revolution, growing enthusiastic over the very noble ideal which was then carrying away the nation. The night of August 4, 1789, when the nobility relinquished their privileges, the Declaration of the Rights of Man and Citizen, the inspiration for which came in part from the ideas expressed in the Declaration of Independence of the American Colonies, the evident necessity of suppressing numerous abuses, the very degree of culture of French society and its philosophical leanings, all serve to explain that eminent minds gave a warm adhesion to the new ideas. Let us not feel surprised, therefore, that men like Laplace, Monge, Meusnier, Lavoisier and many others hastened to proffer their help towards the transformation of society. But the generous ideal of the first days was gradually replaced with suspicion and anger as a result of the resistance encountered, of betrayals and conspiations calling on foreign help to stifle the uprising; thus, was released the reign of

Attendance Sheet of the 6 Nivose, An VI (December 26, 1797) of the Institut National

Terror which brought down the head of Lavoisier. Anything that more or less bore the mark of the "ancien régime" came under suspicion and was destroyed. The academies and, in particular, the *Académie des Sciences* were suppressed. The universities met with the same fate, but they were, in no way, to be regretted from the standpoint of science. Theology had enslaved and sterilized them. They never had been like those in Italy, Germany, or Holland, centers of scientific work. The reaction against the reign of Terror came about swiftly. The suppression of the academies, ordered by the Convention on August 8, 1793, did not last long. They soon were brought back to life in the *Institut National des Sciences et des Arts,* created by the Constitution of the 5th Fructidor, an III (August 22, 1795). For the greater part, the several sections of the *Institut* and in particular, its scientific section, were composed of former academicians. Unfortunately, there were missing the victims of the upheaval, Lavoisier, Condorcet, who poisoned himself in jail to avoid the scaffold, the astronomer Bailly and the président de Saron, also an astronomer, who, before marching to the guillotine, spent the last hours of his life in calculating the orbit of a comet which Meissier had just discovered. Still we see again as members of the new organization: Lagrange, Laplace, Legendre, Monge, Lalande, Méchain, Delambre, Berthollet, Coulomb, Borda, Haüy, Darcet, Fourcroy, Lamarck, A. L. de Jussieu, Desfontaines, Daubenton, Thouin, Bougainville. The continuity was re-established. Bonaparte, victorious in Italy, was proud to be elected to the section of Mechanical Arts, on the 5th Nivose, an VI (December 25, 1797) before he became Consul and Emperor.[1] The *Jardin du Roi,* under suspicion for a time, had sur-

---

[1] In his letter of thanks, he wrote that the real conquests are those made over ignorance.

vived, undergoing a happy transformation, and became the *Muséum National d'Histoire naturelle,* as it remains to-day.   It numbered, or was going to number, among its professors: Lamarck, Cuvier, Et. Geoffroy Saint-Hilaire, Thouin, Desfontaines, Fourcroy, Faujas de Saint-Fond.

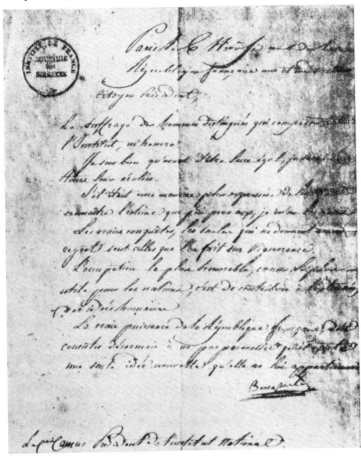

Letter of Thanks of Bonaparte After His Election to the Class of Sciences (Section of Mechanical Arts) of the Institut National, the 5 Nivose, An VI (December 25, 1797)

The Convention, a violent and often implacable but bold and constructive assembly, did not merely destroy; it also created. Among its members, was Lakanal (1762-1845), a former Father of the Oratory, who had a passion for intelligence and institutions which enable it to manifest itself. As much as he could, he was the savior of scientific establishments. He did save the *Jardin du Roi,* tried his best to save the *Académies,* and was one of the most important organizers of the *Institut,* in 1795. "Mon but," écrivait-il plus tard, "était de sauver mon pays, en défendant la cause des lettres . . . de sauver les sciences et ceux qui les honoraient . . . de combattre le vandalisme en provoquant l'établissement d'institutions consacrées à l'instruction publique."[1] In spite of the eminence of the services he rendered, he was to be later, with Monge and Lazare Carnot, among those whom the Restoration deprived of their offices, crossed off the list of the *Institut* and forced into exile. It was in the United States that he took refuge and awaited the hour of his return to France, in 1837.

In summing up the work of the Convention, one must not forget to mention the creation of the great scientific schools which have been a nursery for men of science in France, during the nineteenth century: the *École Polytechnique,* the *École Normale* and establishments such as the *Conservatoire des Arts et Métiers,* the *Bureau des Longitudes.* For some of these creations, the Convention drew its inspiration from institutions of the "ancien régime," like the military schools. Finally, we must put to the credit of the scientific work of the Revolution, the establishment of the metric system and the great geodetic works which this reform brought about.

---

[1] "My aim," he wrote later, "was to save my country, in defending the cause of letters . . . to save sciences and those who honored them . . . to fight vandalism in promoting the establishment of institutions dedicated to public instruction."

During this period, science was associated with great undertakings. The most characteristic episode was the expedition to Egypt. In 1798, Bonaparte sailed forth, taking along with him not only a military staff but also a group of first-class scientists: Monge, Fourier, Berthollet, Geoffroy Saint-Hilaire, Savigny. As soon as he had conquered the Mamelukes, he founded with these men of science, the *Institut d'Egypte* with the object of studying the scientific and historical treasures of the country; the crop they harvested was magnificent.

So we see that, on the whole, the Revolution did not do much harm to French science. It tempered itself through hardships, rejuvenating and extending its institutions. And therefore, as a result, the first decades of the new century were going to witness a splendid blossoming of men and discoveries and constitute, perhaps, the golden age and most brilliant epoch of scientific life in France.

March 10, 1933.

Third Lecture

FIRST DECADES OF THE NINETEENTH CENTURY
(1800-1830)

Mathematics: Cauchy; Évariste Galois.—Physics: Coulomb; Malus and the polarization of light.— Fresnel and the undulatory theory of light.— Ampère and the creation of electrodynamics.— Sadi Carnot and the principles of thermodynamics. —The advances in chemistry.—Natural sciences: Lamarck, Et. Geoffroy Saint-Hilaire; G. Cuvier.

We now stand on the threshold of the nineteenth century; the revolutionary storm had subsided, and, on the whole, given new strength to science by rejuvenating and strengthening political institutions. This state of things was going to determine, from then onwards, the conditions of scientific production.

The creation of the *École Polytechnique* (1794) provided, for several generations, a unique nursery of men of science; this school drew to itself, in a methodical way, the most gifted minds, through an intense mathematical preparation, practising an automatic selection over the whole country. During the first three quarters of the century, almost all the innovators in science came from that school.

The *École Normale* which, in 1795, was organized on too vast a plan, had but an ephemeral existence; but it was reorganized during the Empire, in 1808, in view of securing a constant supply of instructors for secondary schools. For a long time, its existence remained rather precarious on account of the suspicions which the government and especially the Church entertained towards the too liberal teaching reverberating from its halls. However, from 1830 onwards, political conditions became more favorable to the school and its prosperity increased. From the scientific point of view, its recruiting was closely correlated to that of the *École Polytechnique:* same programs, same candidates; the former institution was merely a derivation of the latter; gradually, during the second half of the century, an ever increasing tendency drew former students of the *École Normale* toward research. The most illustrious of them, Pasteur, as sub-director of the school, about 1860, exercised his great influence to intensify and regularize the flow of scientific vocations; meanwhile, on the contrary, more and more attractive opportunities offered to engineers by the economic advance, tended to rarefy vocations of this kind in the *École Polytechnique.*

Such were the two major homes of scientific culture in France, during the greater part of the nineteenth century. Outside of them, there had been created by the Empire fairly numerous faculties of sciences, but, deliberately, they had been planned in a rather narrow fashion; their principal function was to furnish boards of examiners. Cuvier, who filled important administrative positions and therein displayed the brilliant qualities of his genius, as early as 1830, wished to build up in France real, well equipped universities, enjoying a high degree of freedom, on the model of those "German and Italian scientific republics which educated travellers visit with so much pleas-

ure."[1] One may believe that had he not suddenly
died in 1832, carried off by acute myelitis, his fecund
idea might have prevailed, under the liberal régime
of Louis-Philippe's monarchy. The faculties of
sciences, devoid of equipped laboratories, continued
to lead a vegetative life, well into the second half of the
century.

At the end of the reign of Napoléon III, higher
education received a first semblance of organization,
on a wide basis and with an aim to scientific research,
at the hands of a great minister, Duruy; however, the
undertaking was not actually promoted until later than
1870, during the Third Republic; it was not till the
end of the century that real universities were, at last,
constituted, under the impulse of Louis Liard, but not
without opposition. All through the nineteenth century, great minds like Pasteur, Claude Bernard, Wurtz,
had asked for better laboratories, invoking the example
of German universities. The response to such exhorta-
tions is of very recent date and still not quite sufficient.

Such have been the general conditions of the scientific environment in France, since the beginning of
the nineteenth century. It is necessary to know them to
understand the rate of advance of scientific development.

*

* *

At the beginning of the century, a magnificent
growth took place, in all fields, as a result of the in-
heritance from the eighteenth century. Some of the
great masters of the preceding period, whose careers
were drawing to a close, discovered and encouraged
beginners of fruitful promise; thus, a new brilliant

---

[1] On the part played by Cuvier, in the organization of public
instruction under the Empire and the Restoration, see an interesting
article by J. Poirier: *Georges Cuvier, second fondateur de l'Univer-
sité, Revue de Paris, 1er juillet,* 1932.

phalanx sprang forth. If one peruses the list of members of the *Académie*, from 1800 to 1830, one finds by the side of Lagrange, who died in 1813, and Laplace, who passed away only in 1827, after playing to the end a considerable and beneficent rôle, mathematicians and physicists such as Biot, Ampère, Poinsot, Fourier Cauchy, Poisson, Malus, Fresnel, Dulong; astronomers such as Delambre, Arago, Bouvard; chemists such as Fourcroy, Guyton de Morveau, Berthollet and Vauquelin, (contemporaries of Lavoisier), Gay-Lussac, Proust, Chevreul, and Thénard; mineralogists and geologists such as Haüy, Alexandre Brongniart; naturalists such as A. L. de Jussieu, Lamarck, Cuvier, Et. Geoffroy Saint-Hilaire, Latreille, Savigny, de Blainville, to name only the most illustrious in the several branches. Paris was the center which attracted foreigners, and fundamental discoveries were repeatedly going to spring forth.

We are going to bring under review the first period of the nineteenth century, proceeding, of course, rapidly and stopping about 1830, exactly a century ago; we shall reserve for our last three lectures the account of the last hundred years.

*
* *

Mathematics and physics are more and more closely related. Higher mathematics are, indeed, indispensable to the study of physical phenomena, not only to formulate their theory, but also in the interpretation and prevision of experimental results. Conversely, the study of physical phenomena, rather often, suggests the creation of new mathematical theories. A proof of this close interrelation is found in the fact that several of the great men I have just mentioned earned their fame as physicists as much as mathematicians; so did Fourier, Biot, Poisson. This had also happened, by the way, in the preceding century, and Newton, furnished the greatest illustration of it. The name of Fourier has remained connected mostly with the study of the propagation of heat; the analytical methods he created in view of this work have since been applied to many problems in physics; but it is to be noticed that the results obtained were due to his introduction of trigonometrical series, now called the series of Fourier, one of the most important acquisitions in the field of pure mathematics. The English geometrician, Hamilton, ranks Fourier, among mathematicians, at the same level as Lagrange and Laplace, if not above. Laplace himself did not owe his fame only to his work in celestial mechanics and mathematics proper; we already saw that he collaborated with Lavoisier in calorimetry. In the period now to be considered, he tackled numerous questions of physics: capillarity, magnetism, optics, the velocity of sound, etc. . . . The work of Poisson is that of a mathematician almost as much as of a physicist. Ampère, an overflowing genius, also was equally at home in mathematics and physics. The vigor of French science, in that epoch and since then, resided in the generality and depth of mathematical culture.

This being said, we are practically obliged in our outline to take up separately the examination of these two sciences, and we shall begin with pure mathematics.

Mathematics.

There should first be mentioned the mathematical physicists we have just spoken of: Poisson, (1780-1840), a direct disciple of Laplace and often his continuator; Biot (1774-1862), Ampère (1775-1836) and Fourier (1768-1830). The last mentioned was a man of action as much as a man of study; before the

Revolution, he had entered a religious order, the Benedictines; in 1798, he accompanied Bonaparte to Egypt and organized the *Institut d'Égypte* and directed the publication of its works. In 1806, he was a professor at the *École Polytechnique;* then, the Emperor made him an administrator, appointing him prefect of Isère. In this capacity, Fourier showed superior talent in the execution of public works; meanwhile, he established his analytical theory of heat. On account of his fidelity to Napoléon the Restoration removed him from office and refused at first to ratify his election to the *Académie des Sciences.*

Among the mathematicians of the beginning of the nineteenth century in France, two names stand out above the others: Cauchy and Galois.

Cauchy.

Augustin Cauchy (1789-1857) led a life of unrest, although he devoted himself exclusively to mathematical speculations. Endowed with an intransigent conscience, he was an ardent royalist and Catholic; while still quite young, he became a professor at the *École Polytechnique* and at the *Collège de France;* in 1816, at the age of twenty-seven, he was made a member of the *Académie des Sciences;* in 1830, he refused to take the oath to the July Monarchy and expatriated himself; he first went to Turin, the birthplace of Lagrange, then joined King·Charles X and educated the king's grandson, the Count of Chambord. Cauchy came back to France at the end of his life, Napoléon III excusing him from taking the oath. His mathematical work is immense; it has been collected under the auspices of the *Académie des Sciences* and amounts to no less than 27 volumes in 4°; it covers the whole field of mathematics. The center of his work is the theory of the functions of imaginary variables which renovated the field of analysis. Cauchy also made important contributions to mathematical physics, especially in optics.

Évariste Galois.

As for Évariste Galois (1811-1832) his life was impetuous, brief and tragic. Born in 1811, this unruly child of genius, ever since his days at the *lycée*, was seized with a passion for mathematics, but he went through an irregular course of study. While he prepared for the *École Polytechnique*, his teacher detected his genius. Galois, though, failed in the entrance examination, very likely because his paper was too personal and was not understood by the examiners. This looks to us as the first blow of that fate which was to hound him. In 1829, his first memoir was written on the subject of periodic continuous fractions. Cauchy lost the manuscript which he had charge to report. Galois failed again to qualify for the *École Polytechnique*. Disillusioned, he reluctantly entered the *École Normale*, the existence of which was then rather precarious, owing to governmental suspicions. In 1830, he presented another memoir to the *Académie*. This time Fourier, perpetual secretary, died while the manuscript was still in his possession; it never was found among his papers. Of a fiery and bitter turn of mind, Galois threw himself into politics and was expelled from the *École Normale* as a sequence to some trouble. Poisson declared incomprehensible a memoir Galois had submitted to him. Later, seditious speech caused Galois's arrest and imprisonment. He was released, on probation, in March, 1832. In a letter to his friend, Chevalier, he wrote: "Il me manque pour être un savant de n'être que cela. Le cœur en moi se révolte contre la tête."[1]

---

[1] "That which I lack to be a man of science is to be that alone. My heart revolts against my head."

A love affair led him into a duel.   In the night that preceded the encounter, May 29, 1832, he feverishly summed up the researches he was working on.   It is from this letter written to his friend, A. Chevalier, that we know the important results he had arrived at, and the trends of his thought.[1]   On May 30, 1832, at the duel, a bullet hit him in the abdomen and he died the next day, at the hospital, from peritonitis.

His body was taken to the common grave.   The manuscripts he left were published only in 1846 by Liouville, not without difficulty.   Galois' work contributed to the theory of equations data which renewed it by connecting it to the theory of forms and groups.   Conceived and written by a prodigiously gifted young man, these hasty works did not reach a lucid and polished form, but at a distance, from the

Evariste Galois

remoteness which gives their true proportions to things, the most authoritative mathematicians have all proclaimed the incomparable import and profound originality of Galois' work, which includes, in all, about sixty pages.

---

[1] In another still unpublished letter, written at the same time, one finds the beautiful thought used as the epigraph of this book: "Science is the work of the human mind which is destined to study rather than to know, to search for truth rather than to find it." (Quoted from M. E. Picard, *Histoire de la Nation française,* T. xiv, p. xviii.)

The celebrated Norwegian mathematician, Sophus Lie, who made one of the most important contributions to the study of the theory of groups, ranks Galois by the side of Gauss, Abel and Cauchy, among the four greatest mathematicians of the first part of the nineteenth century. Here is part of a judgment of M. Émile Picard[1] in his introduction to the new edition of Galois's works: "Il semble que Galois ait fait en analyse des découvertes au moins aussi importantes (que dans la théorie des équations). . . . On acquiert la conviction qu'il était en possession des résultats les plus essentiels sur les intégrales abéliennes que Riemann devait obtenir vingt-cinq ans plus tard. Si quelques années de plus lui avaient été données pour développer ses idées, il aurait été le glorieux continuateur d'Abel et aurait édifié les parties essentielles de la théorie des fonctions algébriques, telle que nous la connaissons aujourd'hui. Les méditations de Galois portèrent encore plus loin: il termine sa lettre (à Auguste Chevalier) en parlant de l'application à l'analyse transcendante de la théorie de l'ambiguite. On devine à peu près ce qu'il entend par là et, sur ce terrain, qui, comme il le dit, est immense, il reste encore aujourd'hui bien des découvertes à faire.

"Ce n'est pas sans émotion que l'on achève la lecture du testament scientifique de ce jeune homme de 20 ans, écrit la veille du jour où il devait disparaître dans une obscure querelle. Sa mort fut pour la science une perte immense: l'influence de Galois, s'il eût vécu, aurait grandement modifié l'orientation des recherches mathématiques dans notre pays. Je ne me risquerai pas à des comparaisons périlleuses. Galois a sans doute des égaux parmi les grands mathématiciens de

---

[1] *Œuvres mathématiques d'Évariste Galois, rééditées par la Société mathématique de France,* Paris (Gauthier-Villars), 1897 and *Revue générale des Sciences,* T. 8, 1897, p. 339-340.

ce siècle. Aucun ne le surpasse par l'originalité et la profondeur de ses conceptions."[1]

Physics.

Brilliant as may have been mathematical discoveries in the beginning of the nineteenth century, only mathematicians themselves are able to measure their real value. The achievements which, at that time, renewed several major departments of physics are easier to appraise. This science was still completely under the dominance of Newton's conceptions. Laplace, Poisson or Biot were doing their best, in formulating the theories of newly discovered phenomena, to show evidence of elementary forces similar to gravitation. A significant case had been offered in electrostatics, as a result of Coulomb's work, at the end of the eighteenth century.

---

[1] "It seems that Galois made in analysis discoveries at least as important (as in the theory of equations). . . One becomes convinced that he was in possession of the most essential results on Abelian integrals which Riemann was to obtain twenty-five years later. If a few more years had been given him to develop his ideas, he would have been a glorious successor to Abel and would have built up the essential parts of the theory of algebraic functions, as we know it to-day. The meditations of Galois carried even further; he ends his letter (to Auguste Chevalier) by speaking of the application to transcendental analysis of the theory of ambiguity. One can approximately guess what he means by that, and in this field, which, as he says, is immense, there remains even to-day many discoveries to be made.

"It is not without emotion that one finishes reading the scientific will of this young man of twenty, written the day before he was to die in an obscure quarrel. His death was an immense loss for science; the influence of Galois, if he had lived, would have greatly modified the orientation of mathematical research in our country. I will not risk perilous comparisons. There is no doubt that Galois has equals among the great mathematicians of this century. None, however, surpasses him in originality and profundity of conceptions." *

*All known documents on Galois have been utilized by P. Dupuy: *Vie Évariste Galois, Annales de l'École Normale Supérieure*, T. 13, 1896.

Coulomb.

Coulomb (1736-1806) also came from the *Ecole de Mézières* and had been an officer in the engineers. On account of his career, he had been led to deal with various questions of mechanics. To these researches of a practical nature, we owe his works on friction and the laws applying to it; researches in hydraulics led him later to study the viscosity of fluids. But his principal claim to fame was his experimental study of the laws of attraction and repulsion of electrified bodies. The means he used to measure the forces in action was the measuring of the torsion of a wire from which hung the electrified body under observation. He had already applied this method in the study of compasses and magnetism. The tool thus created was the torsion balance. To these researches we also owe the study of the laws of torsion itself. In regard to electrical attractions, Coulomb found that the forces in action, like gravitation itself, diminished in inverse ratio to the square of the distance. The mathematical theory of these phenomena was undertaken by Laplace and completed by Poisson. Likewise, Laplace, in studying phenomena of capillarity, arrived at a law of the Newtonian type.

In optics, Newton's conceptions and his theory of emission ruled supreme. With their help, one tried to explain all newly found optical phenomena, but such explanations, we must say, led to new complexities in the mathematical theory and to far-fetched subtleties.

In experimental physics, Newton had made the fundamental discovery of the decomposition of white light and of the phenomenon of interference, made visible by the formation of colored bands on thin plates; he had recognized in it a periodical phenomenon which is, by the way, the real nature of light. But it was only in the beginning of the nineteenth century that the importance of this character of periodicity was really recognized in the work of the English physician and physicist, Thomas Young. Young had an original and brilliant mind whose activity was dispersed over the most varied subjects; philology, archæology, physiology, etc. . . . ; his works left, in each of these fields, an interesting mark. In physics, he lacked the solid mathematical culture indispensable to bring to completion the study of problems. Between 1801 and 1807, Young published a series of works in which one finds the germs of all the data which were to revolutionize optics on the basis of the wave-theory of light, conceived in the seventeenth century by Huygens. The value of these works was recognized in France, and Young was elected a correspondent of the *Académie* in 1818, and a foreign associate in 1827.

Malus and the Polarization of Light.

The first of the great discoveries in optics at the beginning of the century, was that of the polarization of light made by Malus, in 1808. Malus (1775-1812) was an officer in the engineers corps. In 1792, at the age of 17, he had been admitted to the *École du Génie* at Mézières, which was closed almost immediately afterwards. In 1795, he attended the courses of the newly created *École Polytechnique*. He had taken part in the expedition to Egypt and ever since had been interested in optics. Elected a member of the *Académie des Sciences* in 1810, he died prematurely in 1812, at the age of thirty-seven. In the course of observations on Icelandic spar, which he made in his room by looking through the crystal at the sun reflected by the windows of the Luxembourg, he found that the image disappeared in certain positions. He recognized that this phenomenon was peculiar to reflected light and therefrom, was able to reproduce it at will, under various conditions. This new phenomenon, the polarization of light, stirred up at once much research, notably on the part of Biot and Arago. Almost immediately there were discovered new aspects of it, chromatic polarization, rotating polarization which is caused by certain crystals and also by some dissolutions. Biot endeavored to explain mathematically these properties of light by the Newtonian theory of emission, but only succeeded in formulating very involved and purely verbal explanations.

Fresnel and the Undulatory Theory of Light.

Then appeared Augustin Fresnel (1788-1827) whose capital work was to be built up in the course of a very short life. After his passage through the

Augustin Fresnel

École Polytechnique, Fresnel became a young engineer in the Department of Public Works; he had a dislike for the life and occupations that were his, as an official in the small town of Nyons (Drôme). He filled his leisures by cogitating about optics and particularly about the polarization of light which Malus had just discovered. His hostility to the re-establishment of the Empire, in 1815, caused him to be suspended from his functions and he retired to his mother's home, in the small village of Mathieu, in Calvados. Fortunately, going through Paris, he established relations with Arago. At Mathieu, in studying the rectilinear propagation of light and the phenomena of diffraction which restrict it, he was led to study the question of interferences and tried to explain them in taking periodicity as a basis for the whole theory of light.

With the sole help of the village locksmith, he built a very simple apparatus which proved sufficient to perform the crucial experiments his reasoning had suggested. A few months later, as early as October, 1815, he sent to the *Académie* remarkable communications on diffraction. Thanks to the influence of Arago, he was enabled to reside in Paris, temporarily at first, in 1816, then permanently, beginning with 1818. The problem of diffraction being selected as the subject of a prize by the *Académie,* in 1819, Fresnel sent in a memoir on the question; in an attempt to generalize, he compared the explanations one could give of optical phenomena, either by Newton's corpuscular theory of emission, or by Huygens's undulatory theory which he developed mathematically. The latter offered him the solution of all difficulties in various cases. The memoir of Fresnel was submitted to a commission composed of Laplace, Biot, and Poisson, all three firm believers in the emission theory, and of Arago and Gay-Lussac. This examination turned to the glory of Fresnel under conditions which have often been told, but deserve to be repeated here. Poisson perceived a consequence of the theory which Fresnel had not foreseen, and which was contrary to common sense. Fresnel's theory led to admit that there should be a vividly brilliant spot at the center of the shadow caused by a small opaque disc lighted up by a luminous point. Fresnel, on being told of this objection by Arago, made the experiment. It showed the correctness of Poisson's prevision. Therein lay a striking confirmation of Fresnel's theory, whose memoir was crowned by the *Académie.* From then onwards, in the course of a few years, Fresnel succeeded in giving simple and complete explanations of all known optical phenomena, including polarization, in taking for granted that light results from waves, the vibrations of which are transverse to the direction of propagation. In the case of polarization, these vibrations remain in a fixed plane; in ordinary light, they radiate in all directions. Fresnel brought around to his views almost all former believers in the emission theory and was elected to the *Académie* in 1823. The wave-theory of light, as he constituted and mathematically developed it, ruled in optics, without reservations, all through the nineteenth century.

Fresnel, when elected to the *Académie,* was only thirty-five years old. Unfortunately, his health was gravely impaired and forbade him intensive work. Appointed in the Department of Public Works to the commission on lighthouses, he radically modified the construction of the lights in use and very much increased their power by the invention of the lens belt which is still in use. This invention, which has rendered such great service, was his last piece of work. He died in 1827, at the age of thirty-nine. His amazingly fruitful activity hardly extended beyond 12 years. This was enough for him to completely renew optics; besides, through his conception of the transverse vibrations of light, his work was to have a momentous influence on other parts of physics and mathematics.

Ampère and the Creation of Electrodynamics.

The same years during which Fresnel was doing his work also witnessed, in France, the birth of a new science, electrodynamics, which was in the nineteenth century, and still is to-day, of unequaled importance, both in theoretical science and practical life.

In the eighteenth century, we saw the first and feeble rise of the science of electricity, still limited to the field of electrostatics. Electricity was then produced only by friction on non-conductors or by induction on insulated conductors; in both cases, the discharge came through sparks. The laws of these phenomena had been experimentally established by Coulomb, from 1785 to 1789, by means of the torsion balance. But, interesting as these studies may have been, electrostatics, from the experimental point of view, did not lead later on to considerable advances nor applications. Quite different results were attained in electrodynamics, the science of electric currents flowing through conductors. Electricity accumulated on a body in a static condition causes, at the moment of discharge, an instantaneous current, which, however, on account of its extremely short duration, cannot be studied. The study of currents, therefore, could not begin before the invention of the pile, by the Italian physicist Volta, in 1800. Bonaparte, as First Consul, paid a solemn tribute to this great invention by inviting Volta to come to Paris and repeat his experiments before the *Académie des Sciences* and by showering him with honors. The applications of the pile were, at first, made mostly in England, where the great chemist Davy isolated by the means of a powerful battery the two alkaline metals, potassium and sodium, of which Lavoisier had foreseen the existence; Davy also discovered the electric arc. For this important work, Davy, in spite of the state of war between France and England, received from the *Institut* the great award granted by the First Consul.

In 1819, the Danish physicist, Oersted, in turn, made the unexpected discovery that the passage of the current flowing from the pile through a conducting wire deflected a magnetized needle placed in the vicinity of the wire, the needle tending to take a position perpendicular to the wire. This caused great astonishment and served as a starting point to the researches of Laplace, Savart, Biot and, above all, of Ampère, who was going, in a few weeks, to build up the science of electrodynamics.

Ampère.

André-Marie Ampère (1775-1836) is, at the same time, one of the greatest and most attractive personalities of science.[1] Born in Lyons, brought up in the village of Poleymieux, where his father, an admirer of J. J. Rousseau, was his first and almost only teacher, forming his son's mind without any compulsion, Ampère successively developed a passionate interest in the most divers sciences of his time, as well as in literature, philosophy and poetry. He pursued all those studies by himself. The Revolution suddenly upset his life. His father was guillotined at Lyons, in 1793.

After a long period of complete prostration, Ampère pulled himself together, earned a living, at first, by giving private lessons in mathematics and physics, then became a professor at the *lycées* of Bourg and Lyons. In spite of isolation and material difficulties, his thought was active. A remarkable memoir which he published in 1802, on problems of probability, in reference to games of chance, drew the attention of the astronomers Lalande and Delambre. Soon after, he lived in Paris, first as a *"répétiteur,"* then as a professor at the *École Polytechnique;* in a few years, his works in mathematics brought about his election to the *Académie des Sciences* (1814). Alone, they would suffice to put him in the ranks of the greatest mathematicians of the beginning of the century. But we shall see later that, during this same period, he was formulating capital views on chemistry and the

---

[1] On the most interesting personality of Ampère and the events of his life, see the book of M. L. de Launay: *Le grand Ampère,* Paris *(Perrin),* 1925.

molecular theory of gases. However, his chief claim
to glory was to come out from the experiment through
which the Danish physi-
cist, Oersted, in 1819,
discovered the action of
electric current on a magnetized needle. This ex-
periment became known
in Paris only on Septem-
ber 4, 1820, when Arago,
who had witnessed it in
Geneva, in de la Rive's
laboratory, reproduced
it himself before the
Académie. Through an

A.-M. Ampère

intuition of genius, Ampère, in a few days, identified
magnetism and electricity. He worked in a fever of
which his memoirs and his letters to his son reveal
the ardor.[1] From September 18th to November 6th,

---

[1] Here is a passage from his memoirs which will give us an
idea of this creative fever, as well as of the speed and depth of
his insight. It concerns his first paper, the one he wrote for the
meeting of the *Académie* on September 18: *"Je décrivis les instruments que je me proposais de faire construire et, entre autres,
des spirales et des hélices galvaniques. J'annonçai que ces der-
nières produiraient, dans tous les cas, les mêmes effets que les
aimants. J'entrai ensuite dans quelques détails sur la manière dont
je conçois les aimants, comme devant uniquement leurs propriétés
à des courants électriques circulant dans des plans perpendicu-
laires à leur axe et sur les courants semblables que j'admets dans
le globe terrestre, de sorte que je réduisis tous les phénomènes
magnétiques à des effets purement électriques."* One may see,
in some degree, the flash of genius suddenly illuminating the new
world of electromagnetism.

* "I described the instruments which I proposed to have built,
and among others, galvanic spirals and helices. I announced that
these last would produce in every case the same effects as magnets.
Afterwards, I entered into some details about the manner in which
I conceived of magnets as deriving their properties solely from
electric currents flowing on planes perpendicular to their axis and
about similar currents, the existence of which I admit in the terrestrial globe, so that I reduced all magnetic phenomena to purely
electrical effects."

from week to week, his communications followed one another regularly and fixed the laws of electromagnetism. The crucial experiments which established them were made in his apartment, the only laboratory at his disposal, with apparatus built with his own hands, on a table which is to-day a precious relic preserved at the *Collège de France*. Like many others of the greatest creators of French science, Ampère brought forth his capital discoveries in a penury verging almost on misery. From these was born, in the course of a few weeks, a complete, new and fundamental science, the most fertile in its applications. Already Ampère derived from it the solenoid, the electromagnet, the principle of the telegraph, and a few years later he inspired the machine builder, Pixii, with the idea of the first electromagnetic machine.

Linking, without delay, mathematical study with experimentation, Ampère presented a synthesis of his researches in a *Mémoire sur la théorie des phénomènes électrodynamiques uniquement déduite de l'expérience* which appeared in 1827, in the *Mémoires de l'Académie*. To appreciate the import of Ampère's work, it is enough to quote J. C. Maxwell, whose genius established the electromagnetic theory of light: "The researches by which Ampère established the laws of the mechanical forces electric currents exert on each other stand out among the most brilliant achievements ever reached in science." Maxwell calls Ampère the Newton of electricity.

Another capital discovery, made almost immediately after, completed those of Ampère and perfected that marvelous tool, electrodynamics; this was the discovery of induction. It was the work of an Englishman, Faraday, and was made in 1831; it led to the dynamo and this is enough said to measure its importance. Later on, besides, the discovery of induction led to considerable consequences in the general theoretical conception of magnetic and electrical actions; it prepared the work of Maxwell, which is the preface to all contemporary physics.

Sadi Carnot and the Principles of Thermodynamics.

In the period under review, French physics produced a third achievement, the results of which have been no less important. This time, the work was in the field of heat. On this subject, a series of discoveries, the consequences of which developed but gradually, were made in France in the first twenty years of the new century. We have already noticed in passing, at the end of the eighteenth century, the works of Lavoisier and Laplace, who founded calorimetry, in defining more clearly the notion of quantity of heat, and those of Fourier, which resulted, in 1818, in his *Théorie analytique de la chaleur*. The physical study of gases was to introduce fruitful data and ideas. Gay-Lussac (1778-1850) equally renowned as a chemist and a physicist, in 1806, made the surprising observation that all gases dilate according to the same law; they have the same coefficient of dilatation; on the other hand, Gay-Lussac established that there is always a simple ratio between the volumes of combining gases, as well as between such volumes and that of the gaseous compound obtained. An Italian man of science, Avogadro, in 1813, deduced from this the all-important consequence that, at the same pressure and temperature, equal volumes of all gases contain the same number of molecules. This idea has now been verified by various experimental methods, and one has been able to measure with sufficient accuracy the number of these molecules; this is called the Avogadro's number. Let it be said, in passing, that Ampère, without knowing Avogadro's work, had drawn the same conclusion in a memoir published in 1814.

These several laws inspired a number of researches on the calorimetry of gases, of which it is impossible to give here a detailed account; as a result, Delaroche and Bérard, in 1813, determined the specific heats of gases; besides, two distinct specific heats were defined, one for a constant volume, the other for a constant pressure; these latter determinations were due to Clément and Desormes in 1819. The new notions thus acquired later played an important part; at that time, they enabled to solve a discrepancy which had proved a stumblingblock, the discrepancy between the velocity of sound in the air as measured experimentally and that which Newton had determined through calculations. Lagrange and Euler had vainly tried to explain it. Laplace and Poisson met with success by taking into account caloric phenomena which accompany the propagation of the sound-wave and applying to them the data provided by Clément and Desormes. New determinations of the velocity of sound, worked out in 1832, on the model of those made in 1738, then showed a perfect agreement between the experimental result and that reached through the formulas now accurate, established by Laplace, in 1816.

At the same time, a solitary study was being pursued which did not attract attention, but was later to become of capital importance, as the basis of a new and essential branch of physics, the one that connects mechanics to caloric phenomena (thermodynamics). It was the work of Sadi Carnot (1796-1832).

Sadi Carnot was one of the sons of Lazare Carnot (1753-1823), a minister of War under the Convention; Lazare Carnot became famous for his political rôle, when France was invaded by the armies of the European kings; he earned the distinction of being called the "Organizer of Victory." At the same time, Lazare Carnot was an eminent man of science. Graduating from the *École de Mézières,* he held a commission in the engineers before the Revolution, and left remarkable works in mechanics and calculus; his scientific glory, though, was absorbed into the radiance of, his son's. After the downfall of the Empire, in 1815, Lazare Carnot was deprived of all his official functions by the Restoration; like Monge, he was crossed off the list of members of the *Institut* in 1816, and died as an exile, in 1823, at Magdeburg.

Sadi Carnot, born in 1796, entered the *École Polytechnique* in 1812, at the age of sixteen; he graduated as an officer in the engineers, but did not remain long in the army, under the Restoration, preferring to devote himself to scientific meditations. The epidemic of cholera which devastated Paris in 1832 carried him away at the age of thirty-six. The only thing he had published was a pamphlet of about a hundred pages, which appeared in 1924, under the title: *Réflexions*

Sadi Carnot

sur la puissance motrice du feu et les moyens propres à développer cette puissance. This had been printed at his own cost, in a limited edition, and had not been communicated to any scientific society or periodical. It undoubtedly remained unnoticed partly on account of its originality. It was, indeed, outside the scientific preoccupations of the time, and some of the phenomena relating to the ideas Carnot expounded had remained unexplained. Carnot's work, above all, bore the mark of an engineer's interest; it was the cogitation of a deep, solitary mind and thereby original. Lord Kelvin (Sir William Thomson), one of the greatest names in British science, on a visit to Paris, in 1845, had great difficulty in securing Carnot's pamphlet.

The fundamental concern of Carnot, an engineer's concern, was to establish in a precise manner the relation between heat and motive power, a problem suggested by the study of the steam engine, the industrial rise of which was becoming a fact. Carnot was interested in its rational improvement. In his own words, he tried to establish the theory of "Toute machine à feu imaginable, quelle que soit la substance mise en œuvre et quelle que soit la manière dont on agisse sur elle."[1]

Thus, in 1824, he succeeded in formulating the principle with which his name is still linked: "La puissance motrice de la chaleur est indépendante des agents mis en œuvre pour la réaliser; sa quantité est fixée uniquement par les températures des corps entre lesquels se fait, en dernière analyse, le transport de calorique."[2] At that time, he still thought of the caloric as a particular entity, a special fluid, in the same way it had generally been conceived until then.

This pamphlet, so full of new ideas, might have been lost for all time if, ten years later, it had not attracted the attention of an engineer and mathematician Clapeyron (1799-1864), who, after Carnot's death, in 1834, in a famous memoir, *Sur la puissance du feu,* revealed its worth and novelty and developed its ideas, giving them the form and symbols still in

---

[1] "Any conceivable fire engine, whatever fuel be put to work and in whatever way worked upon."

[2] "The motive power of heat is independent of agents set to work to produce it; its quantity is only determined by the temperatures of the bodies between which the transfer of caloric takes place in the last analysis."

use. Lord Kelvin in England, Clausius in Germany,
were to give Carnot's principle its definitive and fundamental place in science. It has become, in the logical development of thermodynamics, the second law of
this science, the first being that of the mechanical
equivalence of work and heat.

It is a fact that Carnot had also perceived and formulated this first principle. Some of his papers which
escaped being destroyed after his death, yielded an
unpublished memoir in which the principle of equivalence is clearly set down as the following words show:
"D'après quelques idées que je me suis formées sur
la théorie de la chaleur, la production d'une unité de
puissance motrice nécessite la destruction de 2.70 unités
de chaleur."[1]   His notes also contain the program of
experiments he probably would have conducted, if he
had not lacked time to make accurate measurements
of this equivalence; they have been carried out later
by physicists who were not aware of Carnot's inten-
tions. Finally, these posthumous writings contain the
obvious proof that he had given up the notion of
caloric conceived as a special fluid; indeed, he wrote:
"La chaleur n'est autre chose que la puissance motrice
ou plutôt le mouvement qui a changé de forme. C'est
un mouvement dans les particules des corps. Partout
où il y a destruction de puissance motrice il y a, en
même temps, production de chaleur en quantité pré-
cisément proportionnelle. Réciproquement, partout où
il y a destruction de chaleur, il y a en même temps

---

[1] "From some ideas I have formed on the theory of heat, the production of one unit of motive power requires the destruction of
2.70 units of heat." *

* One has been able to repeat the computations by which Carnot
reached that evaluation. The figure 2.70 is equivalent in the language of our days to 370 kilogrammetres per calory. The exact
value of the mechanical equivalent of heat, as determined later by
various methods, is 425.

production de puissance motrice."[1] Therefore, these records peremptorily prove that death did not give him time enough to make these conceptions publicly known. They were to be reached a few years later, in 1843, by Rob Mayer[2] in Germany and Joule in England, who both followed exactly the way outlined by Carnot. But he had, indeed, conceived and established them, so that, as a whole, the basis of thermodynamics truly was a creation of his genius. To appraise the value of his discoveries, let us be content to cite this judgment of Lord Kelvin: "In the entire field of science, there is nothing greater than the work of Sadi Carnot."

Thus, between 1820 and 1830, in the field of physics, and to limit ourselves to capital facts, French science had reformed optics with Fresnel, created electrodynamics with Ampère and thermodynamics with Carnot.

*
* *

The Advance in Chemistry.

Chemistry, during the same period, made manifest the fruitful consequences of Lavoisier's work. Going beyond the scope of particular facts, the works of

[1] "Heat is nothing else but motive power, or rather motion which has undergone a change of form. It is a motion in the particles of bodies. Wherever there is destruction of motive power, there is, at the same time, production of heat in a directly proportional quantity. In a reciprocal way, wherever there is destruction of heat, there is, at the same time, production of motive power." *

* Sadi Carnot—*Biographie et manuscrit publiés sous les auspices de l'Académie des Sciences avec une introduction de M. Émile Picard.* Paris, (Gauthier-Villars), 1927, p. 81 et fac-simile du manuscrit, feuillet 7, recto, p. 44.

[2] Marc Seguin, the inventor of the tubular boiler, from which dates the modern locomotive, also formulated definite views on the equivalence of work and heat, and the principle of the conservation of energy, as early as 1839.

Berthollet (1748-1822) and Proust (1754-1826) established the general laws of definite and multiple proportions in chemical compounds; the works of Gay-Lussac (1778-1850), in 1805, resulted in determining the laws of volumes in gaseous combinations; they brought into relief the uniformity of the mode of dilatation of gases; from those laws, Avogadro and Ampère drew the consequences we have already pointed out in relation to the number of molecules contained in a certain volume. So, a considerable part of the general laws of chemistry was established by French men of science of the period now under review.

As for specific results, the new methods and ideas led to a number of important discoveries and to the isolation of new elements. In England, Davy, with the help of the pile, isolated the alkaline metals. In France, Gay-Lussac and Thénard (1777-1857) soon secured potassium and sodium in greater quantities by another method. They also showed that chlorine is a simple element, not a compound as was formerly believed. They isolated bore, in 1808; Vauquelin isolated chromium, in 1797; Courtois, iodine in 1811, and Balard (1802-1876), bromine, in 1826. In 1815, Gay-Lussac prepared cyanogene, which is a compound of carbon and nitrogen, but acts like a simple body; it was the first discovered of these bodies, later called *radicals,* which play a capital part in organic chemistry. In the field of organic chemistry, then almost untouched, the name of Chevreul (1786-1889) stands out; he died over a hundred years old, but was then already the leading authority in the chemistry of immediate principles, derived from animal or vegetable substances. From 1813 to 1823, his great work was the chemical study of fats and their general division in fat acids and glycerin. Therefrom, considerable industrial advances were evolved, particularly in the manufacture of soaps and stearic candles; this brought a remarkable improvement in house lighting conditions. Pharmaceutical chemists, studying the immediate principles in plants, extracted from them a whole new class of organic bases, alkaloids. The first, solanine, was isolated by a chemist from Franche-Comté, Desfosses. In this direction, Pelletier and Caventou especially became famous by isolating, from 1817 to 1822, strychnine, veratrin, brucine and quinine, this last one being of major importance as a remedy against malaria. It was also in France that then appeared the first data relative to the mechanism of chemical transformations inside of organisms. In 1822, Dubrunfaut showed how to saccharify feculas by the action of germinated barley; Payen and Persoz isolated from it, that is from malt, *diastasis,* the prototype of soluble ferments which are the essential agents in the mass of chemical actions taking place inside the living cell. Finally, J. B. Dumas (1800-1884) revealed himself at that time; his first works received widespread recognition. He was one of the first exponents of the notions of function and substitution, two fundamental conceptions in the development of organic chemistry. He rapidly acquired a very great authority of which, by the way, he did not always make the best use later on, as we shall see from the account of the next period. In a more detailed study, it would be fitting to mention other chemical discoveries made at that time, but this would go beyond the scope of these lectures.

*
* *

Natural Sciences.

The advance in natural sciences in France, at the beginning of the nineteenth century, was no less brilliant than in the other departments of science. But, for the present, even at the cost of retracing my steps later on, I shall leave aside physiology, since its full growth took place only after 1830.

At the end of the eighteenth century, there were, in France, good naturalists, among whom the chief ones had been contributors to the *Encyclopédie Méthodique,* undertaken by the editor Panckouke, in 1780. But we do not need to dwell on them. Let us pass immediately to three great men whom one may consider the founders of modern morphology: Lamarck, Étienne Geoffroy Saint-Hilaire and Georges Cuvier.

Lamarck (1744-1829) had a twofold career. Until 1794, when he reached the age of fifty, he was an eminent botanist, the author of a *Flore française* which enjoyed a great vogue; he was a close friend of Buffon, whose influence on his ideas is undeniable. On the reorganization of the *Jardin du Roi* in 1793, he was entrusted with the professorship of Invertebrata; this led to his becoming a zoölogist for the last thirty-five years of his life. It is, indeed, remarkable that his mind was plastic enough to permit of his veering around towards another science in which he did work of considerable importance. The "animals without vertebra," our modern invertebrata, formed the most confused part of the animal kingdom. Thanks to Lamarck, their classification made remarkable advances in the detail of which, of course, I cannot venture here. To-day, the name of Lamarck survives much more as that of the founder of the evolutionary doctrines, which he worked out during the last years of the eighteenth century and expressed in the famous laws of his *Philosophie Zoologique,* published in 1809. While the general conceptions of Lamarck had little influence during his lifetime, they acquired much after Darwin, and, in a way, they remain to-day, one of the poles of biological thought.

Framed at a time when accurate knowledge was lacking on most points, they present a much too simple solution of a tremendously complex problem. According to these views, we should think of organisms as incessantly modifying and adapting themselves under the direct action of environment through a process of individual variations, which, henceforth, heredity would transmit to succeeding generations. Accurate

Lamarck

experimentation has shown, in recent times, that this is not verified in Nature, at least within the scope and time limits of our ability to experiment. In any case, adaptation cannot be taken for granted with the same degree of generality which Lamarck assigned to it; the intrinsic characters of the organisms certainly have on evolution a more momentous action. But, though the present trend of ideas be unfavorable to Lamarckism, I hardly think it possible to radically set it aside. At any rate, whatever may be the fate in store for the doctrine, its part in the development of biological and evolutionary thought will have been considerable. Let us say, in passing, that Lamarck, in 1802, invented the word *biologie,* at the same time as the German naturalist, Treviranus. An encyclopedic mind, formed in the philosophical environment of the eighteenth century, Lamarck held on the various sciences very broad views, some venturesome and obsolete, others in advance of their time; such were, for instance, those he formulated in geology on present phenomena and their value in explaining the past.

Étienne Geoffroy Saint-Hilaire.

Étienne Geoffroy Saint-Hilaire (1772-1844) was only twenty-two years old when, on the recommendation of the mineralogist, Haüy, at the time of the reorganization of the *Jardin du Roi,* he was entrusted with the chair of Superior Animals. A few years later, he was found among the group of men of science who accompanied Bonaparte to Egypt; this gave him an opportunity for numerous works of high interest. One of the first occasions on which he exerted his influence, in a very generous as well as fortunate way, presented itself almost immediately after he took up his work at the *Muséum:* Geoffroy was instrumental in bringing to Paris young Georges Cuvier, who was hardly older than himself; Cuvier was then tutoring in the d'Héricy family, in Normandy, where he had been devoting himself to anatomical work on marine animals; these works had attracted some attention. Geoffroy and Cuvier, both young and full of ardor, formed a strong friendship and started on a fruitful collaboration. Inspired by the works of Jussieu, they applied to animals the principle of the subordination of characters, and brought about, in a few years, great progress in comparative anatomy. The opposite tendencies of their minds gradually led them into diverging paths and, in 1830, finally caused them to clash, in a well-known controversy. Cuvier and Geoffroy, together or separately, discovered and formulated the most important general principles of comparative anatomy. One may say that they founded this science. Geoffroy, for his part, applied himself more to the living and fossil mammals and reptiles. He deduced from his numerous anatomical researches, the essential principle of connections, according to which an organ can hypertrophy, transform itself, or atrophy, but not change its position in relation to the rest of the organism. This principle is one of the most general and safest guides in comparative anatomy. He was the first to perceive the importance of embryogeny in the interpretation and comparison of the structures of adult animals; he also was the first to understand all the significance of rudimentary organs, as traces of anterior functional mechanisms. Finally, he was the founder of teratology. The several kinds of results we have just mentioned, naturally, led him to accept evolutionary views based on comparative anatomy and to favor Lamarck's ideas; he never agreed, however, with all of the latter's conclusions. Geoffroy Saint-Hilaire must, therefore, be counted among the founders of the transformist doctrines. Being of a philosophic turn of mind and fond of generalizing, he let himself be guided and even carried away by ideas; this is why he was not always able to limit the scope of his hypotheses and deductions. Thus, gradually, the substance of his *Philosophie anatomique* was summed up in his theory of unity of plan in organic composition, in the animal kingdom; this was indeed a magnificent conception, but unduly simplified, as he imagined it; when he tried to prove it, he was led to entirely artificial and untenable interpretations and comparisons. Cuvier, therefore, came out against it on many sound grounds, but exaggerated the value of the opposite facts.

Cuvier.

Through his perspicuity, his ability to group positive facts and present them in a brilliant synthesis, the amazing wealth of significant facts he was able to discover and through his rôle as an inspirer, Georges

Cuvier (1769-1832) almost at once earned an authority, which, in the course of his career as a man of science, became more and more pronounced, not only in France but in the whole scientific world. During his lifetime he exerted a real supremacy in zoölogy; from every country, scientists flocked to him; the informal receptions he held, each week, were the meeting place of scholars from all over Europe; the naturalist, Quoy, has preserved their remembrance in writing.[1] The influence of Cuvier's ideas survived him a long time; during the greater part of the nineteenth century, his collaborators and direct or indirect disciples kept it alive in France, in too tyrannical a fashion, often with less broad-mindedness than the master himself would have shown. Cuvier's own work is so vast that one cannot think of analyzing it; one can only define its main lines. As he himself said, he wished to establish the study of invertebrata on an accurate knowledge of their anatomy, a knowledge which was almost entirely lacking before him. Through his own efforts and those of his collaborators, the task was carried out in remarkable fashion, considering the times. Outside of his numerous memoirs on specific questions, the synthesis of Cuvier's investigations was presented in his *Leçons d'Anatomie comparée* and in *Le Règne animal distribué d'après son organisation.* These works were the basis of all further research, in the first half of the nineteenth century. Undoubtedly, the classification which Cuvier achieved bears to-day the mark of many corrections. The four fundamental types of organization, which he distinguished and regarded as absolute realities, no longer have the same value; however, we must go back to his own time and compare his works with what existed before

---

[1] *Notes intimes sur Georges Cuvier,* rédigées par le **Dr. Quoy,** pour son ami J. Desjardins de Maurice. **Bull. Scient. France Belgique,** T. 41, 1907, p. 459.

him. His general conceptions corresponded to a zoölogical philosophy which is now antiquated, in its major part. Cuvier had remained steadfastly faithful to the idea of the fixity of the species and to creationism. To him, the animal kingdom was what the thought of the Creator had made it; the mission of the zoölogist was to find out its plan, taking as a basis principles disclosed by careful study, such as the principles of subordination of characters, of correlation, and of living conditions.

Georges Cuvier

The organization of any animal form was exactly the one designed by the Creator to make its life possible under conditions determined by a rigid general teleology. This interpretation of Nature was altogether opposed to that of Lamarck and Geoffroy Saint-Hilaire; it is indefensible nowadays. Whatever may be said of such philosophical views, Cuvier showed a sagacity of the highest order in his anatomic investigations of the several types of animals. His mastery particularly asserted itself in the study of fossil animals, mere fragments being often the only available data, and especially in the study of tertiary mammals. His *Recherches sur les ossements fossiles* is worth the admiration which has been bestowed upon it; while it did not create palæontology in full, at any rate it brought forth the first advance of this science. At the beginning of this work is found the famous *Discours sur les Révolutions du Globe,* in which Cuvier, who prided himself on his strict adherence to positive facts, went beyond them in amazing fashion, and reached conclusions which the progress of geology soon ruined. Nevertheless, in spite of the reservations the general doctrine calls for, Cuvier's work remains as a magnificent evidence of very great intellectual powers. Among the biologists and men of science of modern times, he still retains a place in the foremost ranks.

He surrounded himself with a score of collaborators such as Blainville, Duvernoy, Duméril, Latreille, Valenciennes, d'Orbigny, Laurillard and others; they helped him considerably in the detail of his anatomical researches; after his death, they formed a school, which, as we shall see, remained too obstinately attached to the general conceptions of their master.

Other names, which fade out in the radiance of the three great protagonists we have just evoked, would deserve to be recalled; the names of Cuvier's collaborators whom I have mentioned above, of Savigny, a companion to Geoffroy Saint-Hilaire in the expedition to Egypt, and whose scientific career was checked at an early date by ill health; the names of Victor Audouin and Antoine Dugès, who both died young, etc.

*

* *

To complete the account of the activity of French science in the field of natural sciences, during this brilliant epoch, one must not forget that the tradition of the great voyages, born in the eighteenth century, was maintained under the Consulate, the Empire, and the Restoration; the scope of these scientific explorations was even extended, and they brought back, especially from the antarctic seas, a considerable amount of material rich in novelty. It would be too long to list all those expeditions, their programs or their results. Their published records remain as documents of great worth, even to-day. I shall limit myself to recall the names of the naval officers in command, and of the naturalists who took part in them; among the former: Baudin, de Freycinet, Duperrey, Dumont d'Urville; among the latter: Péron, Lesueur, Bory de Saint-Vincent, Quoy, Gaimard, Lesson, etc. All those names are familiar to zoölogists who study marine animals.

*
* *

Thus, from one side to the other of the scientific horizon, from the most abstract conceptions of higher mathematics to the concrete and minute details of descriptive natural history, the first three decades of the nineteenth century were marked, in France, by a profound change in intellectual outlook and by great discoveries.

March 15, 1933.

Fourth Lecture

MIDDLE AND THE END OF THE NINETEENTH CENTURY

(1830-1900)

MATHEMATICS, PHYSICS, CHEMISTRY.

Positive science: Auguste Comte.—Mathematics.—
Astronomy: Le Verrier.—Physics: Regnault,
Fizeau and Foucault.—Electricity: The applications: The transmission of force, the refrigerating industry, photography, . etc.—Chemistry:
mineral chemistry.—Organic chemistry: Laurent
and Gerhardt, founders of the atomic theory.—
H. Sainte-Claire Deville and physical chemistry
dissociation.—Mineralogy.

Our study, in the preceding lecture, of the period
from 1800 to 1830 has brought us to within a century of the present. It remains for us, therefore, to
take an inventory of the last hundred years in the
three lectures that follow. This task is a more difficult one, because the sedimentation of facts dealing
with more recent years has not yet been definitely
established. It is accordingly a more delicate matter
to distinguish those facts which are really important

and more embarrassing, too, to speak of scientists who are still alive. I shall reserve the last lecture for a study of the contemporary period, to which we shall assign the last years of the nineteenth century as a point of departure. Naturally, this date has been chosen arbitrarily and is, in any case, only approximate. Consequently, the interval from 1830 to 1900 will be the subject of this lecture, which will deal with mathematics, physics, and chemistry, and of the following, which will be reserved for the biological sciences.

Positive Science—Auguste Comte.

In the forefront of this period (1830-1900) we shall place the figure of a man who was not a great scientific creator, but who faithfully expressed certain characteristics of the scientific mentality of the French, and in particular that of the nineteenth century. We are referring to the founder of positivism, Auguste Comte (1798-1857), a man who had the profound culture of the mathematician and who had surveyed the field of knowledge of his times in the manner of a philosopher. His thought certainly reflects that of many contemporary scientists. When it had been expressed in the Cours de Philosophie positive[1], it exercised in turn an undeniable influence, in some ways beneficent, in others inhibitory. Naturally, I am disregarding the purely philosophical aspect of Comte's work as well as the tendency it gradually developed towards the end of his life to realize a substitute for religious mysticism and to form the basis of a scientifically grounded moral philosophy. I shall consider it only in so far as it attempts to define and delimit science itself.

---

[1] 6 volumes (1830-1842).

Auguste Comte had been trained at the *École Polytechnique* and for a great many years performed the duties of "répétiteur" and examiner in mathematics at that institution. He constructed his philosophical system from the scientific point of view. Its basis is a natural classification of the sciences, expressing their genealogy, their historical development and their progressive concatenation, in order of increasing complexity. It goes from astronomy and mathematics to the social sciences, passing through the physical, chemical and biological sciences. Upon this idea, which is in a sense historic, and as a whole quite consistent with reality, is superimposed the law of the *three states,* i. e. that all knowledge passes through three phases, the theological, the metaphysical and finally the positive. This last state is that of science strictly speaking, which no longer seeks to penetrate the very essence of things, since that is of a metaphysical nature, but limits itself to what can really be known, i. e. to phenomena studied in their relations, and the conditions of their succession and dependence. It is not preoccupied with their essential substratum, which is inaccessible to us. The field, as thus defined, is that of positive science, to which the scientist must limit himself, if he would work with certainty and clearness.

This general conception has certainly exercised considerable influence in France. At least, it has found itself in agreement with the intellectual tendencies of the majority of French men of science, invariably skeptical with regard to metaphysics, and ever concerned with the exactness and lucidity of the proof. This proceeds from the application of the Cartesian principle to science, nor is it from this very moment that this tendency has manifested itself. Réaumur is an "esprit positif." Cuvier also lays claim to a positive mind and professes to confine himself to facts. This tendency has in a general way been a contributing factor in discarding all that was suspected of being based upon vague and unverifiable conceptions, upon which limitless speculations and explanations which have only a purely verbal, more or less dissimulated value can sustain themselves. We see here a very characteristic difference in the attitudes of French and German science, especially in the province of biology. As we go along, we shall find this tendency reappearing on many occasions. In certain cases, it has doubtless had the disadvantage of imposing too rigid a restraint upon the imagination, of diverting the mind from conceptions which, although themselves unverifiable, have, nevertheless, served as precious guides in new fields. Moreover, certain of these conceptions or *working hypotheses,* as the Germans say, have later become positive realities in the science of the morrow. The exaggeration of the positivist doctrine has perhaps clipped the wings of French scientists, at times, in the second half of the nineteenth century, in the physical and chemical sciences and also in biology.

Mathematics.

Mathematics, at least, escape the usurpation of the positivist conception. They belong in an eminent degree to the province of imagination. They create entirely fictitious beings, which, when once created, are ruled according to the laws of strict logic. Besides, it has been found that these creations often vividly illuminate certain aspects of reality in the physical world, by means of the properties thus combined.

The splendor of French mathematics, under the influence of conditions of education which we have already explained, was prolonged and intensified during the entire nineteenth century. The *École Polytechnique* and the *École Normale* exercised a great attraction over young people. Thus an automatic and

efficacious selection of minds highly gifted for the
exact sciences was assured. One might even ask
whether there was not a certain abuse of the mathe
matical faculty, developing to an excessive degree this
penchant for logical exactness, and forcing it to sub-
tlety. Mathematics are the indispensable tool of research in physics, but the field of physics is the real,
with its complexity, which can only be attained by ap-
proximations; and the latter require of necessity an
attenuation of pure, rigorous logic. Reality cannot be
contained in its totality in the limited and restrictive
conditions which the mathematician is obliged to impose upon himself in order to establish his equations.
Besides, as the physical world came to be better known,
it escaped from the rigidity of the mathematical frame,
which at first served to explore it. In this respect, a
distinct evolution was realized in the course of the
nineteenth century. At this time, that is to say about
1830, the prestige of the mathematical work of Laplace, Fourier, Poisson, etc., held sovereign sway. They
had reduced physical phenomena, studied in terms of
their infinitesimal structure, to precise mathematical
laws, of the same type, in general, as that of Newtonian
attraction. Such was the case for magnetism, electrostatics, capillarity. Contrariwise, on the strength of
experience, the notion will gradually assert itself that
such rigid simplicity as that formerly assumed, tallies
only with observation considered as a whole and prac-
ticed more or less on the scale of our sense organs:
apparent simplicity results from compensation between
elementary phenomena, which, individually, escape rigorous determinism. As we shall see, most physical
laws no longer appear to-day as absolute and verified
on any scale and under all conditions, but as pure *statistical* laws, perfectly accurate, in fact, only for a
specified scale, that which our senses enable us to
attain directly; on an inferior scale, when elementary

actions remain undetermined, though compensating themselves through great numbers, these laws become evanescent. Even so, elementary actions may still be submitted to mathematical analysis, but under new forms mostly derived from the theory of probability. To this infinitely complex reality, the somewhat traditional spirit of mathematics is not perfectly adequate.

If I might be permitted to evoke some memories of my youth, I would refer to the state of mind that prevailed forty years ago at the *École Normale*. In this limited but very representative little world, mathematics constituted an intellectual aristocracy. There reigned a kind of snobbery of scientific rigor which was not apt to favor the orientation of beginning careers toward physics, and still less towards chemistry. With respect to the latter, many affected an attitude which the term "scorn" would scarcely suffice to indicate. There prevailed at that time in the environment certain conditions to which one can perhaps attribute the rather paradoxical fact that the very great development of mathematics in France, at the end of the last century, was not accompanied by a corresponding blossoming in the field of theoretical physics.

The world of modern mathematics is practically inaccessible to the uninitiated and even to those who, having applied themselves to it, more or less profoundly, in the time of their youth, have subsequently turned to other sciences. I experience a real scruple, not indeed in formulating conclusions on my own account, as I would not allow myself to do so, but in merely evoking the personality and the work of the mathematicians, whom the opinion of their peers designates for our admiration. I consider myself as a mere echo, in citing here a certain number of men and works.

The development of the French school of mathematics in the nineteenth century was uninterrupted.

We have already given prominence to the name of Cauchy, who carried on a considerable production until 1857. Poisson, as much a mathematician as a physicist, who died in 1840, was succeeded, so to speak, by Lamé (1793-1877) in the questions that the former had studied. A brilliant series of geometricians, in the strict sense of the word, enlarged and diversified the field of pure geometry: Poncelet (1788-1867), taken prisoner in the course of the Russian campaign in 1812, devoted the forced leisure of his captivity in the prisons of Saratof to mathematics, and created projective geometry. Michel Chasles (1793-1880), renewed and extended considerably the field of pure geometry. Gaston Darboux (1842-1917), applying all the resources of analysis to the study of surfaces, developed infinitesimal geometry to a remarkable degree. To these names, others must be added, such as those of Liouville (1809-1882) and Laguerre (1834-1886), etc.

It is even more difficult to set a limit to the list of illustrious names in the field of analysis. Charles Hermite (1822-1901), while still a student at the *École Polytechnique,* had revealed himself, in letters written to the great German mathematician Jacobi on the subject of the functions of Abel and the elliptical functions. From this field he never strayed during his long, laborious career, but his work has extended just as forcefully to the many branches of analysis, particularly to the theory of numbers. He had maintained all through his life an enthusiasm for mathematics, which made him speak of it as of a reality filled with beauty. Henri Poincaré (1854-1912), whose precocity was no less remarkable, dominated all the branches of mathematics by his indefatigable labor. Thirty volumes and more than 500 memoirs on the most elevated questions were the result of his life, entirely devoted to speculative thought, and prematurely and abruptly cut off, while in the full activity of production. As early as 1881, his discovery of the functions named by him *fuchsian,* which generalize in a

Charles Hermite

direct and complete manner the ellipitical functions, had made him the equal of the greatest mathematicians who preceded him. It has been said of this theory he created that it was "the crowning glory of the principal achievements of the nineteenth century." But one recognizes the profound traces left by the work of Poincaré in nearly every branch of mathematics: algebra, the theory of analytical functions, functions defined by differential equations, the theory of groups, the theory of numbers, the theory of wholes (*ensembles*), the problem of the three bodies, geometry, mathematical physics—and we shall see what decisive value certain suggestions that he made have had for experimental physics. While one generally ranks the names of Hermite and Poincaré as unrivaled, those who have left behind them a remarkable work are numerous: Liouville, V. Puiseux (1820-1883), Joseph Bertrand (1822-1900), Camille Jordan (1838-1922), G. Halphen (1844-1889), G. Humbert (1859-1921), P. Appell (1855-1930), etc.

In the field of mechanics, one must add to the preceding list of names those of Poinsot (1779-1859), Coriolis (1797-1843), Saint-Venant (1797-1886), and

Boussinesq (1842-1929), whose work in hydronamics has been rich in profound innovations.

Astronomy.

The history of astronomy and of the sciences which are related to it, such as celestial mechanics and geodesy, during the nineteenth century, would be worthy of being treated at length. First, we would find again the names of several of the mathematicians whom we have just mentioned. Above all we would meet that of H. Poincaré, who, by introducing the new resources of analysis into the problem of celestial mechanics, has profoundly modified them and whose comprehensive work on *Les Méthodes nouvelles de la Mécanique céleste* has, according to

Henri Poincaré

the authorized assertion of M. B. Baillaud, renewed the foundations of this science. Among the astronomers, strictly speaking, we shall recall the names of Delaunay (1816-1872), who devoted a large part of his life to the study of the movements of the moon, F. Tisserand (1845-1896) and H. Andoyer (1862-1929).

The most spectacular event in astronomy in the nineteenth century was certainly the discovery of the planet Neptune by U. Le Verrier (1811-1877), in 1846. An engineer in the state manufactories and a former pupil of the *École Polytechnique,* like Fresnel, le Verrier had, at first, devoted his leisure hours to the study of celestial mechanics. He undertook, at the suggestion of Arago, to explain the irregularities in the movements of the planet Uranus, upon the basis of the body of laws established by Laplace. He attributed these irregularities to the influence of a more distant disturbing planet. From this hypothesis, he calculated the mass and the position of the assumed star, and, having completed his calculations, he wrote on the 18th of September, 1846, to the astronomer Galle, of the Berlin Observatory, who had in his possession a detailed chart of that region of the sky where the sought star should be located. On the very day of the receipt of the letter, the 23rd of September, Galle found the prophesied planet, only 52 minutes removed from the position calculated by Le Verrier. This discovery, which justly inspired the admiration of his contemporaries, was the most brilliant imaginable confirmation of the laws of celestial mechanics, the crowning achievement of the work accomplished from Newton to Laplace.[1]

In the course of the nineteenth century, observatories were established in several provincial centers.[2] These, besides the one in Paris, have contributed their share to the progress of positional astronomy and particularly to our knowledge of comets and asteroids.

A new branch of astronomy, the most productive of results in the contemporary period, astrophysics, originated in the nineteenth century, thanks to the progress in physics. Among the founders of this science in France, figures J. Janssen (1824-1907), who

---

[1] An astronomer of the Greenwich Observatory, Adams, had done the same work independently of Le Verrier and announced shortly afterwards that he had arrived at the same results: but the priority of the discovery belongs unquestionably to Le Verrier.

[2] Toulouse, Bordeaux, Marseille, Lyon, Nice, Alger.

organized the Observatory of Meudon in that direction. His efforts have been successfully continued by M. Deslandres. Astrophyics requires, above all, powerful instruments, and it is on that account that the development of this science in France has been hindered.

In geodesy and its applications to cartography France continued in the active part that she had played before. In the eighteenth century we saw her produce the map of Cassini. In the nineteenth, the Army Geographic Bureau, basing its work on new triangulations, first completed the general map of the territory on a scale of 1:80000 and, more recently, more and more precise partial maps on greater scales. These cartographical precision works have been extended since the middle of the nineteenth century to the colonies, beginning with Northern Africa. In geodesy, the French network of triangulation has been united with the English network since 1861 and the Algerian network with the Spanish in 1879. A general and precise triangulation of the French Alps has been undertaken and completed by M. P. Helbronner, with only his private resources. He has, moreover, linked up the triangulation of the Alps with Corsica, the latter being connected with Italy on the other side. Finally, to retake, with modern precision, the eighteenth century measurement of an arc of a meridian in the neighborhood of the equator, France, from 1901 to 1906 sent an important mission under the direction of Commandant (since General) Bourgeois, for the purpose of measuring an arc of more than 5° in the equator.

Let us link to geodesy leveling operations, a field in which again France showed the way, with the general leveling survey of the country, carried out between 1855 and 1867, under the direction of Bourdaloue, an engineer.

Physics.

Physics, as the beginning of the nineteenth century, had passed through a phase of astonishing creative power, as we have seen, and as the names of Fresnel, Ampère and Carnot suffice to attest. The rest of the century, without presenting any discoveries of the same scope, was to witness new and important advances and superb experimental achievements.

Such was the work, first, of Victor Regnault (1810-1878). An assistant to Gay-Lussac after he left the *École Polytechnique,* Regnault began with works on chemistry. But, from 1840 on, he devoted himself, until 1870, when the Franco-German War was to cut short his scientific life, to the study of the properties of vapors and gases and to the precise numerical determination of their various properties. He thus provided physics with a body of data which have had a fundamental importance. Regnault continued and extended the previous work of Dulong and Arago on water vapor and that of Pouillet, who had been the first to establish variations in the compressibility of gases, in connection with the law of Mariotte. Regnault has determined the limits within which the law of Mariotte is exact. He applied to his researches an astonishing experimental ingenuity and precision. Their results have been assembled in three volumes: *Relation des expériences entreprises par ordre du Ministre des Travaux publics pour déterminer les principales lois et les données physiques nécessaires au calcul de la machine à vapeur.*[1] They are full of numerical data, the use of which has been valuable both in pure science and in practice. Regnault was essentially an "esprit positif," unconcerned with theory. He excelled in distinguishing facts, but he was too much their

[1] Account of experiments undertaken by order of the Minister of Public Works to determine the principal laws and the physical data necessary for measurements of the steam engine.

slave and thus remained a stranger to the development of thermodynamics, to which his own works directly related him. And nevertheless, as Berthelot wrote: "Son œuvre a un coté philosophique, sans la connaissance duquel on ne comprendrait ni son rôle, ni l'influence qu'il a exercée. Jusque là, chaque physicien, accoutumé par Laplace et Fourier à la rectitude artificielle des représentations mathématiques, s'efforcait de tirer de ses recherches quelque expression générale qu'il proclamait aussitôt une loi universelle de la nature. Regnault a concouru, plus que personne, à faire disparaître de la science de telles conceptions absolues pour y substituer la notion des relations approximatives, vraies seulement entre certaines limites, au delà desquelles elles se transforment ou s'évanouissent."[1] We accordingly find here a first verification of what I said above on the nature of physical laws and their mathematical representation. The close of Regnault's life was darkened and rendered barren by the misfortunes which the War of 1870 brought. His son, Henri Regnault, a ·painter of great talent, was killed, and, in his laboratory at Sèvres, his papers and apparatus were destroyed by the German troops with skillful barbarity.

One may consider as an extension of the work of Regnault the study of gases at increasing pressures, calculated by hundreds and even by thousands of atmospheres. This study, pursued in various countries,

---

[1] "His work has a philosophical side, without the knowledge of which one could understand neither the part he played nor the influence which he exercised. Until then, every physicist, accustomed by Laplace and Fourier to the artificial correctness of mathematical representations, tried to infer from his researches some general expression which he immediately proclaimed a universal law of nature. Regnault contributed, more than anyone, to the disappearance of such absolute conceptions from science. He substituted for them the notion of approximate relationships, true only within certain limits, beyond which they change or disappear."
Quoted from P. Langevin, *La physique au Collège de France, 1932.*

soon led to the liquefaction of the supposedly permanent gases, accomplished in France in 1878 by L. Cailletet (1832-1913) with small volumes. The study of the properties of gases at these great pressures was also the subject of persevering and important researches on the part of Amagat (1841-1919). Then by the alternate application of pressure and expansion, the liquefaction of gases entered the domain of industry and the name of M. Georges Claude is at present in its foreground.

Fizeau and Foucault.

Among the physicists of the middle of the nineteenth century, two most remarkable experimentalists were Hippolyte Fizeau (1819-1896) and Léon Foucault (1819-1868). Exact contemporaries, they were at first collaborators, and they have this in common, to distinguish them from the general run of physicists in their time, that they were self-taught men and had come to physics from outside the pale of the usual mathematical formation. Both attained a brilliant originality and their works almost always opened important new roads. They had both begun by studies in medicine, and the subject of their first researches was photography, which had just been discovered by Daguerre. They were the first to obtain good photographs of the surface of the sun.

Encouraged by Arago, whom we see on many an occasion, outside of his personal researches, acting the fruitful part of inspirer, they undertook separately and achieved almost simultaneously, by altogether different methods, the direct measurement of the velocity of light. They succeeded in determining it, in spite of its enormous speed (300,000 kilometers a second) over a limited space; Fizeau, by means of his cogwheel method, over some kilometers between Suresnes and Montmartre; Foucault, by means of a revolving mirror, drawn by a little steam turbine at a speed of 400 revolutions a second, measured it inside his laboratory, over a distance of a few meters. The numbers obtained agreed with the conclusion reached in the seventeenth century by Roemer, by means of the eclipses of Jupiter's satellites; but this agreement furnished henceforth a precious means of measuring astronomical distances. The two methods have since been used again with increased precision, that of Fizeau, in France by A. Cornu (1841-1902); that of Foucault, in the United States by Michelson and by the astronomer Newcomb. The velocity of light plays a capital part in contemporary physics. Foucault was able, by means of his method, to measure directly and to compare the velocity of light in two media of different refringence such as air and water. These measurements have shown that it is slower in water than in air, as Fermat maintained against Descartes and the undulatory theory against that of emission. The calculations of Foucault have thus come once more to the support of the work of Fresnel. We owe many other important results to Fizeau and Foucault. They applied the production and the study of interferences to measurements of extreme precision in the expansion of bodies, in the study of the form of surfaces, in astronomical computations. Interferences have thus come into common use in the workshop. The wave length has become, to use Fizeau's expression, a natural micrometer, a unit of measurement independent of all systems and escaping all possible destruction. It was thus made clear that the metric system should be connected with this new unit, by measuring the meter in relation to a determined wave length. That has been done at the International Bureau of Weights and Measures, with the aid of Michelson.

It is also to Fizeau that we are indebted for having succeeded in transferring to the field of optics, the phenomenon observed in 1842 by the Austrian physicist Doppler in acoustics, namely that the height of a sound varies according as its source approaches or recedes from the observer.   In optics, as Fizeau showed in 1848, this phenomenon is expressed by a displacement of the fringes of interference.   This fact, when Fizeau first established it, was only a curiosity; verified since then by various physicists, it has acquired to-day a particular interest by furnishing a means of measuring the velocities with which extremely remote stars recede from or come near us.[1]

Foucault, a machinist of marvelous ingenuity, improved to a considerable extent the construction of astronomical instruments and, notably, of telescopes, by putting into effect rational and practical methods of polishing and silver-plating mirrors.   It is he, too, who, in 1851, effected beneath the cupola of the Pantheon in Paris, a striking demonstration of the rotation of the earth about its axis, by means of a pendulum 67 meters long, whose plane of oscillation turned 360° in 24 hours.   Proceeding from this conception, he invented the gyroscope at the suggestion of the geometrician Poinsot.   To-day, it is put to various applications and, in particular, furnishes the northerly direction, without observation of the stars or magnetic compass: the gyroscopic compasses are being substituted little by little for magnetic compasses on steel ships and especially on board submarines.

Electricity.

In the field of electricity, the French physicists of the nineteenth century have made numerous and important contributions.   In 1834, Pouillet (1790-1868)

---

[1] The use of this method with nebula results in a displacement of the fringes, which would indicate that all these stars recede from us with enormous velocity.   One can ask oneself whether the observed phenomenon may not be due to another cause, heretofore unknown.

succeeded in measuring in an exact way electrical resistances, the laws of which had just been established by the German physicist Ohm. Thermo-electricity has been the work of Favre and Silbermann, between 1850 and 1855, of Antoine Becquerel (1788-1878), inventor of the two-fluid, constant voltage pile, and of his son Edmond Becquerel (1820-1891). Fizeau and Foucault introduced into the construction of apparatus with inductive currents improvements which resulted in the construction of Ruhmkorff's solenoid, built in Paris, in 1855. Also in Paris, a Belgian workman, Z. Gramme, contrived the first dynamo, in 1869.[1] Likewise, it was in France that galvanometers were mostly perfected and that, in particular, Deprez and d'Arsonval invented the moving coil galvanometer, now in universal use. Another instrument, currently used in physics laboratories, the capillary electrometer, was devised by C. Lippmann (1845-1921), as a prelude to a brilliant career as a physicist. Finally, it was in Paris that the first International Congress of Electricity assembled in 1881; this congress definitely established the values of electrical units named after Ohm, Volta, Ampère, Coulomb and Faraday. One of the inspirers of the congress was E. Mascart (1837-1903), a physicist, who, besides doing important personal work, was among the best to understand and follow the advances of science during his time.[2]

---

[1] In 1860, was born another French invention, the electrical accumulator devised by G. Planté, an amateur physicist. Its full development did not take place until after the invention of the dynamo which enables to charge it.

[2] We may also recall here that another international congress held in Paris, in 1872, had founded the International Bureau of Standards (Bureau International des Poids et Mesures), since located in the Pavillon de Breteuil at Saint-Cloud. This bureau was entrusted with the care, testing and duplication of the standards of the metric system. The highly precise work necessary for the accomplishment of this task has been the point of departure for numerous researches. These have contributed numerous and important improvements to the science of physical measurements, metrology.

The Transmission of Power.

Toward the close of the nineteenth century, the use of a new type of motive power also originated in France. "Water power," or the utilization of the energy furnished by waterfalls, has become increasingly essential to modern industry. The turbine, together with the fundamental organ of this industry, the dynamo, had been invented in 1830 by the French engineer, Fourneyron. And it is through the inspiration of Marcel Déprez (1843-1918), that about 1881, the transmission of power to a distance, by means of electrical conductors and dynamos, began. This is another of the primordial aspects of contemporary industry. The cradle of these important applications of electricity is Grenoble, at the foot of the Alps.

The Refrigerating Industry.

Another industry based upon the data of physics, and whose importance increases every day is the refrigerating industry. The first frigorific machines, which operate through the liquefaction and evaporation of ammonia, are the creation of Charles Tellier. Although, like many other inventors, he has not found in the development of this industry the fortune to which he should have had a right, the preservation of food commodities and their transportation with the help of refrigeration, are a valuable acquisition for which humanity is indebted to Tellier.

Photography.

Of all the applications that sprung from physics in the nineteenth century, none has perhaps been more productive nor more varied than photography. It is truly a new sense in man's service, revealing, recording with faithful exactness, materializing and measuring phenomena, many of which would escape us, were it not for the camera. This wonderful invention is the work of two Frenchmen, who, after some trials carried on independently, had joined their efforts in 1829, in order to complete this invention: Niepce (1765-1833), an amateur, occupied with research and invention in the field of physics during the greater part of his life, and Daguerre (1789-1851), a scene painter who sought to fixate the images produced by the camera-obscura. After the death of Niepce, Daguerre succeeded in obtaining lasting images on metallic plates, the daguerreotypes. The invention was announced at the *Académie des Sciences* by Arago, at the beginning of 1839. The exposure necessary to obtain these first impressions was 15 minutes long, and Arago expressed the hope that, " 'neath the glowing sun of Egypt, the time could be reduced by half." But at the same time, he already perceived the vast services which the new invention could render, and, in order not to hinder its development, he had obtained Daguerre's promise not to withhold the secret of his methods of procedure, in return for a recompense from the state, which the government would assure him. It is due to this initiative that photography immediately entered the public domain. Its applications to science in France, and particularly to astronomy, were immediate, in the hands of Fizeau and Foucault and later the Henry brothers. This has led to the undertaking of the photographic chart of the heavens, a vast international enterprise still in the process of execution, initiated at the Observatory of Paris. Besides, photography is to-day, for ordinary astronomy as for astrophysics, the eye of the astronomer, infinitely more powerful and certain than human vision. The wonderful applications of photography to the study of motion are also, in a large measure, of French origin. From this point of view, the name of the physiologist Marey, which we shall later find again, must first be recalled, and the names of the brothers Auguste and Louis Lumière, of Lyons, who, in 1895, invented the cinematograph, one of the most characteristic tools of contemporary life. Finally, color photography was achieved by various methods conceived in France. As early as 1847, Edmond Becquerel was able to photograph the colors of the spectrum, but did not succeed in fixating them. Twenty years later, Charles Cros and Ducos de Hauron, working independently, solved the problem by superposing three stereotype plates obtained through three screens of suitable colors. This method was made practical in 1896 by the Lumière brothers, who achieved simultaneously the equivalent of the three screens on the plate itself by means of their autochromatic plates. On the other hand, Lippmann obtained in 1894, by an altogether different procedure, by means of interferences of stationary waves, marvelous photographs of the colors, without the intervention of any coloring matter.

Other Various Applications.

One also should add to the credit of French science in the nineteenth century, other applications of the advances in physics. The internal-combustion engine, the basis of the development of the automobile, dirigible, and aëroplane, was invented by Lenoir, in 1860. The dirigible itself was a French creation. The original idea goes back to Meusnier in 1783. In 1872, Dupuy de Lome applied the ideas of Meusnier, But the first effective results were due to a captain of engineers, Ch. Renard, in 1884. On August 9th, for the first time, an oblong balloon, shaped like a cigar, the *France,* impelled by a propeller and an 8 HP. motor, starting from Meudon, came back to its point of departure, after completing a circuit of more than 7 kilometers. Victor Hugo, who was still living, glorified the event in a poem. Modern performances, brought about by the improvement of motors, are but the development of Ch. Renard's work.

We are also indebted to French science and especially to Teisserenc de Bort (1855-1913), for the initial exploration of the higher strata of the atmosphere, by means of trial balloons, and the ascertaining of the peculiarities of the stratosphere.

Finally, besides inventions which were fully developed in France, one must mention those which had their beginning there, although they were brought to completion elsewhere. Thus, before 1830, Ampère, clearly described the principle of the electric telegraph, as an application of his work in electromagnetism; however, he conceived it in a form too complex for practical purposes. In 1854, a modest telegraph operator, Charles Bourseul, was able to transmit sound through electric currents by means of an apparatus not essentially different from our modern telephone, but he did not succeed in transmitting articulate speech. This capital achievement was the work of Graham Bell, twenty years later.

One sees to how many different problems, in the domain of pure science or of its applications, French physicists have, in the course of the nineteenth century, contributed developments of considerable significance and interest.

Chemistry.

The progress of chemistry has become more and more intensified all through the nineteenth century. France has contributed to it a much larger share, in so far as innovating and fruitful ideas are concerned, than has been apparent in the course of time.

For many ideas that were of French origin, have found elsewhere, and particularly in Germany a field more favorable to their development, a collective organization of work, on the one hand, and on the other, well equipped laboratories which have been lacking in France for too long a time. The period that we are considering has been for chemistry, too, a stage of daring and fruitful theoretical conceptions. In regard to these a certain distrust and an excess of the critical attitude have prevailed in France, retarding their expansion in the land of their origin.

All through the nineteenth century, a series of eminent chemists, besides those already cited, has followed in order in France, occupying the great chairs in Paris, sitting at the *Académie des Sciences,* exercising thus, upon the movement of ideas and researches an influence often too exclusive and sometimes, as we shall see, retarding the progress of ideas which were not theirs. Among these masters of chemistry in the middle of the nineteenth century we must name first of all J. B. Dumas (1800-1884), who, having obtained both notoriety and honor while very young, enjoyed a considerable influence. He was one of the first pioneers in organic chemistry and he had broad conceptions about chemistry in general. Besides him, let us recall the names of Thénard (1777-1857), whose works have dealt principally with mineral chemistry, Cahours (1813-1891), Frémy (1814-1894), V. Regnault, who began with chemistry, before going on to physics. Pasteur (1822-1895), himself was first a chemist before becoming a biologist, and his works on tartaric acids would suffice to save his name from oblivion. Boussingault (1802-1887), was the great pioneer in the applications of chemistry to agriculture and in the rational use of fertilizers. Th. Schloesing (1824-1919), and his son, A. T. Schloesing (1856-1930), A. Muntz (1846-1917), P. Dehérain (1830-1902), continued his work. Marcelin Berthelot (1827-1907) is one of the greatest names of chemistry, and his work has encompassed many different fields of it. Sainte-Claire Deville (1818-1881) was also a chemist of the first rank, whose retinue is formed by a series of remarkable pupils: H. Debray (1827-1888), L. Troost (1825-1901), A. Joly, H. Moissan (1852-1907), etc. Both master and pupils limited their efforts to mineral chemistry, to which they made eminent contributions. In organic chemistry, Ad. Wurtz (1817-1884), Ch. Friedel (1832-1899), Ed. Grimaux (1835-1900), Arm. Gautier (1837-1910), A. Haller (1849-1925), P. Schutzenberger (1829-1897), occupied the first rank, by reason of their research and teaching; but above all they carried on the work of two great innovators who died in their youth, Aug. Laurent (1807-1853), and Charles Gerhardt (1816-1856). They were the founders of the atomic theory, to which the progress of all modern organic chemistry is due. Hence, men were not lacking. Let us see what their principal discoveries were.

Mineral Chemistry.

In mineral chemistry, many particular results deserve to be mentioned: the isolation of elements, from that of uranium by Peligot (1811-1890), to that of fluorine by Moissan, the discovery of metals through spectral analysis by Lecoq de Boisbaudran (gallium, samarium), Lamy (thallium), Demarçay (germanium), and Urbain (lutecium). For the classic classification of metalloids we are indebted to J. B. Dumas. In 1862, Bécuyer de Chancourtois, anticipating Mendéléef, had proposed a general classification of the elements, by arranging them along a helix ("vis tellurique"), at different levels, in order of ascending atomic weights. Since the helix was attached to a cylinder, elements of the same family were placed upon the same generatrix of the cylinder, as later in one of the vertical columns in Mendéléef's table. This entire French school of mineral chemists was particu-

Marcelin Berthelot

larly skilled in dealing with high temperatures. H. Sainte-Claire Deville and Debray were the first to constitute aluminum metallurgy. They also effected the fusion of platinum and the separation of all metals of the same family, whose properties they studied. At the close of the century, Moissan, together with his pupils, utilizing the advances which had been made in electricity, renewed our knowledge of many more or less refractory metals and metalloids, and the chemistry of carbides, silicides and borides. He contributed more than any one to the creation of a new industry, electrothermics. By the dry way or by the humid way, Sainte-Claire Deville, de Sénarmont, Verneuil, Ebelmen, and Hautefeuille have reproduced many crystallized natural minerals. In this way, Moissan succeeded in obtaining the diamond. At the same time, the numerous achievements which had been accomplished by the dry way led to more precise methods of evaluating high temperatures by H. Sainte-Claire Deville, Violle (through optical processes), and M. H. le Chatelier (by calorimetrical processes). In the study of the properties of industrial metals, one could not forget the name of Osmond, to whom the progress of metallography is due.

In almost every branch of chemistry, in the discovery of important single facts as with general conceptions, we come across the name of Marcelin Berthelot. In organic chemistry, he was the pioneer in synthesis, and devoted himself particularly to its first terms, from acetylene (C^2H^2), which he obtained directly from hydrogen and carbon, to the most complex compounds. He did not limit himself to conceiving the reactions leading to the production of bodies; he determined to elucidate their energetic mechanism, from which thermochemistry emerged. He drew up the thermic balance sheet of each reaction in order to complete its ponderal balance sheet. The ensemble of his researches in this subject have culminated in two great works: *Essai de Mécanique chimique fondée sur la Thermochimie* (1879), and *Thermochimie, lois et données numériques* (1897). The general methods which he conceived in the course of his researches and, in particular, the use of the calorimetric bomb, enabled him to effect great advances in the chemistry of explosives. M. Vielle was his collaborator in this part of his work.

Organic Chemistry.

Organic chemistry had given rise to important works, from the beginning of the century. We have already encountered those of Chevreul on fats and we saw that J. B. Dumas had made important contributions of facts and doctrines to this field, from the very outset of his career. While mineral chemistry, because of the multiplicity and diversity of its elements, was to remain rather disparate in character, organic chemistry rapidly gave rise to a far more coherent system: that is due to the fact that it is in the main the chemistry of only one element, carbon. Consequently, it has a real unity, in spite of the innumerable combinations to which carbon lends itself. The atomic theory has made it possible to encompass it in its entirety and form one of the finest structures built by modern science—an edifice whose strength, breadth, and general value become increasingly noteworthy.

It is in France that the atomic theory was worked out, in difficult and even painful circumstances which it would be interesting and fitting to recall in some detail.

The first foundations of the atomic theory were laid by the English chemist Dalton, as early as 1809. It is he who revived the Greek conception of atoms as indivisible units of matter, in a specified form. Gay-Lussac's discoveries, on volumetric relations in combinations of gases and on the uniformity of their expansion, had led Avogadro and Ampère, as we have seen, to formulate the idea that all gases, under similar conditions of temperature and pressure, contain, in a given volume, the same number of molecules—these last being the smallest possible parts of a body able to exist independently. Finally, as the component elements replace each other in combinations according to proportional weights (to which the name of "equivalents" or "proportional numbers" was given), the French physicists Dulong and Petit brought to light a constant relationship between proportional numbers and specific temperatures of gases. The celebrated Swedish chemist Berzélius built on this data, about the year 1820, an atomic theory which enjoyed much popularity from 1820 to 1840. In 1826, J. B. Dumas established numerous densities of vapors in order to determine proportional numbers. But he was soon faced with a series of exceptions which were difficult to explain and which made him skeptical about the theory itself. "Si j'étais le maître," he wrote, "j'effacerais le mot atome de la science, persuadé qu'il va plus loin que l'expérience et jamais, en Chimie, nous ne devons aller plus loin que l'expérience."[1]   In general, an obviously true maxim in which one perceives an echo of Aug. Comte's ideas, a wide and presumptive distrust of all that is not tangible reality.   And after all, it was proved later on that the supposed anomalies were due to the fact that no sufficiently clear distinction had been made between the atom and the molecule. Avogadro's law applies to molecules but these are formed of groups of atoms, brought about, in each case, with the condensation of varying volumes.   This is the case even for simple bodies. The union of two atoms of hydrogen is necessary to form a molecule of such a body. If one takes this fact into consideration, the atomic theory becomes an indisputable truth, but with important differences in regard to the conception Berzélius had formulated.

In 1831, there appeared in Paris a chemist who formulated a correct interpretation of Avogadro's law. It was Gaudin, who published a memoir on this subject, a memoir whose importance became apparent later, but which was so little noticed at the time that only the first part was actually published, in *Annales de Chimie*. The second part, although announced, was never published and was perhaps rejected as unfit. From 1830 to 1850, the truth perceived by Gaudin gradually came to light on an experimental basis, in the midst of discussions which were often confused and impassioned with injustice.   It is difficult to imagine to-day that a discussion of a scientific nature can assume the violence that these attained.   Two very young men, complete strangers to one another, whose relationship had its beginning in controversies and opposition before it developed into a very friendly collaboration, Antoine Laurent and Charles Gerhardt, were the champions and originators of the atomic

---

[1] "If I were ruler, I should erase the word atom from science, convinced as I am that it goes beyond experience, and, in chemistry, we should never go further than experience."

theory; but they came into conflict with the tenacious and often spiteful opposition of the majority of chemists who ruled over the world of science. They both wore themselves out in this struggle and died prematurely, before they could see the triumph of their ideas.

Laurent and Gerhardt, little by little, founded the basis of the atomic theory through experimentation, but allowing themselves to be guided largely by hypothesis and answering objections with facts. None the less, they were going counter to established ideas, and were in conflict with the prominent men of science, without always knowing how to spare their opponents. Instead of being helped, therefore, they had to suffer violent opposition. They were both poor; at first, as teachers, Laurent at Bordeaux, Gerhardt at Montpellier, they were unfortunately completely deprived of laboratory equipment. This was at the time, and continued to be for a long while the weak point in French science, resulting from the lack of universities. This poverty in working apparatus brought them both back to Paris in rather precarious circumstances. Nevertheless, they collaborated for a few years in a very productive way. No one did justice to their discoveries—far from it. Laurent could not obtain, in 1850, the chair of chemistry which happened to be vacant at the *Collège de France,* and where he could have worked under good conditions. Gerhardt was obliged to leave Paris once more, to occupy a chair at the University of Strasbourg. Weakened by work and struggle, they were both destined to die prematurely: Laurent was dying of tuberculosis in 1853, at the age of forty-six; in 1856, Gerhardt, forty years old, was carried away within a few days by an attack of pneumonia. It was outside of France that their ideas gained ground most rapidly and had the most fruitful results. At the time of his death, Gerhardt was publishing a treatise on organic chemistry, which he had not had the time to complete. It enjoyed enormous success in various countries, and has had a considerable influence on organic chemists of the second half of the century. It has been a starting point for modern organic chemistry.

It is not possible here to show in detail the importance of Laurent's and Gerhardt's work. Several lectures would be necessary and I should not be qualified to give them. Those who would like to know the various phases of the struggles by means of which these two men of science reformed organic chemistry will find a splendid account of them in the book devoted by Ed. Grimaux[1] to Gerhardt's work. I shall confine myself to calling in evidence a testimony the authority and impartiality of which are unimpeachable, that of one of the great German chemists of the second half of the nineteenth century, Hofmann, who in 1862, when entering the Berlin Academy, wrote that, "it was from Gerhardt's doctrines that he had drawn the most precious stimulation to new investigations, the most precise information for the exact interpretation of the facts observed, and lastly it was to his notation that he owed the simplest expression for explaining and grouping the results obtained." All the modern organic chemists are the disciples of Laurent and Gerhardt. From them dates a new period in the history of their science. Since their death, moreover, they have had a series of eminent continuers in France, who have developed the atomic theory in the most brilliant manner and have drawn from it many important discoveries: Wurtz, a native of Strasbourg like Gerhardt, Schutzenberger, also of Strasbourg, Ch. Friedel, an Alsatian, too, Ed. Grimaux, Armand Gautier, all men of the same generation, which has

---

[1] Ed. Grimaux and Ch. Gerhardt fils: *Charles Gerhardt, sa vie, son œuvre, sa correspondance*—Paris, 1900.

completed the reform work of Laurent and Gerhardt, and which has been followed by a score of younger investigators.

At the death of Gerhardt, there remained to be added to the equipment of the atomic theory an important concept, that of valence; to tell the truth, it resulted indirectly from his concept of types; evolved later, it has permitted the establishment of formulæ of the composition of organic compounds.   Another important idea has subsequently been added to the fabric of the theory, that of the tetrahedral structure and of the asymmetry of the carbon atom.   It proceeds from Pasteur's studies, and it was formulated in 1874 in Holland by van t'Hoff and in France by Le Bel simultaneously.   From that has sprung stereochemistry, in which the formulæ of composition extend to the three dimensions of space.

We can see that the atomic theory, which is to some extent the skeleton of modern organic chemistry, has been, in its essential parts, elaborated in France.   It should, then, have developed there more quickly than in other countries.   Unfortunately, this has not been the case, because even if, immediately after the death of Laurent and Gerhardt there was in France a brilliant school of organic chemists adopting the new ideas, there existed beside it, none the less, a radical and tenacious opposition to the atomic theory; this opposition emanated from men who were eminent and also possessed very great influence, particularly in the field of teaching.   H. Sainte-Claire Deville and Berthelot remained systematic adversaries, which prevented the atomic theory from penetrating elementary instruction until about 1890.   My generation was educated in the *lycées* with the old figures still, and in 1887 at the *École Normale,* I could still see in some of our chemistry instructors attached to the old tradition of Sainte-Claire Deville, a scoffing skepticism with regard to the atomic theory. We even witnessed the paradox of being taught mineral chemistry by equivalents and organic chemistry by atoms.

This opposition, so regrettable on the part of men as eminent as Berthelot and Sainte-Claire Deville, who are in the front rank of the chemists of the entire nineteenth century, is certainly explained to a rather large extent as we have already been able to observe in the case of J. B. Dumas, by the general tendencies of their minds, by the concern for immediate, tangible proof, resulting from experiment itself, and by an instinctive distrust regarding hypothetical interpretations. Similar suspicion was shown a little later also, though in a milder way, in physics, likewise on the subject of atomical conceptions. At the bottom, it is allied with the spirit of the positive philosophy of Aug. Comte, and, in this case, we could not dispute the fact that there has been a factor of sterility.

H. Sainte-Claire Deville and Physical Chemistry.— Dissociation.

With regard to Sainte-Claire Deville, we should not stop at this negative impression. We must not forget the great part he had in the progress of mineral chemistry. But we must point out particularly that he was the initiator of a new branch of chemistry through which it joins physics, *chemi-physics*. For a long time, the two sciences were opposed to each other by the nature of the phenomena that were their subject: the realm of physics was, in a general way, that of reversible, continuous phenomena; that of chemistry, phenomena, such as combinations, that modify the properties of bodies in a discontinuous and irreversible way. Now Sainte-Claire Deville brought these spheres together. In his experiments at high temperature, he saw that bodies perfectly constant under ordinary conditions, bodies such as water, are separated into their constitutive elements, hydrogen and oxygen, at a high temperature, and that this phenomenon is reversible at a lower temperature. This dissociation complies with fixed laws. Under given temperature conditions, a well defined equilibrium is established between the compound and the separate components. The new idea of dissociation has been of major importance in science. Sainte-Claire Deville immediately had the intuition that it was a question of phenomena of a great generality, and, in fact, he extended the knowledge of it to many cases, just as did H. Debray, his pupil. The idea of equilibrium had been perceived under other circumstances by Berthelot, at the beginning of the nineteenth century, in his *Statique chimique* in reference to the mutual displacement of acids and bases in salts. These problems immediately received a wide development at the hands of different chemists, Berthelot and Péan de Saint Gilles (equilibrium in the formation of ethers from alcohols), Isambert, Hautefeuille, Ditte, Lemoine (allotropic transmutations).[1] To these reversible transmutations are applied the laws of thermodynamics, and Carnot's principle especially. The mathematical study of it has been made by Massieu (who discovered and studied the characteristic functions of these transmutations), and by Moutier. The systematic study of it has been developed mainly by M. H. le Chatelier, in his *Recherches expérimentales et théoriques sur les équilibres chimiques* (1888), and that led him to clarify and spread abroad the fundamental studies of J. W. Gibbs. To the same fields of science belong the remarkable investigations of F. Raoult (1830-1901), which have rendered prominent the laws of two cate-

---

[1] Although not immediately connected with these questions, mention should be made here of the studies, as exact as they were elaborate, made by Gernez (1834-1910), on various phenomena such as the allotropic transmutations of sulphur, the superfusion, and supersaturation of certain solutions.

gories of phenomena that also belong to physical chemistry: cryoscopy, that is to say, the study of the lowering of the point of congelation of solutions, and tonometry, that of the raising of the point of volatilization of solvents. The concepts thus established play a fundamental rôle in the theory of solutions and the measure of molecular weights, where these concepts are now in daily use, and they also come within the limits of thermodynamics. In the realm of mathematical physics, all these questions have given rise to the exhaustive studies of P. Duhem (1861-1916).

Mineralogy.

To complete this all too rapid review, a few words remain to be said about a science which, on its descriptive side, is related to the natural sciences, and on its theoretical sides, to physics: Mineralogy. The structure of crystals is one of the branches of physics and, at present, one of the most important. We have seen it originate with Haüy at the end of the eighteenth century. In the nineteenth, it was in France that it made the most decided progress, and that were formulated the concepts which at the present time have assumed a very great importance, at the same time that they have received tangible confirmation. Haüy had understood the crystals as resulting from a piling up of component molecules, the form of which he conceived according to that of microscopic crystals. Bravais (1811-1863), took the view that the molecules in a crystal, instead of being arranged accidentally like amorphous bodies, are placed regularly following directions forming a network whose elements of symmetry are shown by the axes of the crystals themselves. He subjected the properties of such a structure to a mathematical analysis based on the considerations of symmetry. These studies, which, during a half century, had only a purely theoretical value, achieved a distinct, concrete significance about twenty years ago. Bravais' ideas have, indeed, found themselves materialized by the experimental study of the diffraction of X rays through crystals, a study of which the German physicist Laue was the initiator. They had been developed meanwhile and put in a less abstract form by E. Mallard (1833-1894). Outside of the fundamental studies on the general structure of crystals, it would be worth mentioning many mineralogical investigations connected with various problems of physics and chemistry. We have already recalled the reproduction of numerous natural minerals by different chemists or mineralogists. Nor could we omit, with regard to this science, the investigation through which Pasteur opened his career, the discovery of the hemihedrism of tartrates and the relations he observed between this fact, the rotatory force of solutions and the molecular dissymmetry that one was led to associate with it. That is what subsequently brought about the beginning of stereochemistry, as we have already seen.

March 17, 1933.

Fifth Lecture

FRENCH SCIENCE FROM 1830 TO 1900

The Biological Sciences

The descriptive natural sciences: Zoölogy, botany, geology. — Physiology. — Claude Bernard. — The work of Pasteur.

After having reviewed mathematics, physics, and chemistry during the middle and end of the nineteenth century, it remains for us to follow the evolution of the biological sciences during the same period.

The Descriptive Natural Sciences—Zoölogy.

We have seen the brilliant achievements that zoölogy offered from 1800 to 1830, through Lamarck, Geoffroy Saint-Hilaire and Cuvier. Lamarck died in 1829, after a long, wearisome old age. Cuvier died in 1832, carried off abruptly in a few days, in the full activity of creation and in full force. Geoffroy Saint-Hilaire in turn died in 1844. As we have already stated, the influence of Cuvier's ideas continued a long time. It was still felt toward the end of the century. His collaborators and his pupils occupied for a long time the main professorships of the Museum and the Faculty of Sciences at Paris: Blainville, Valenciennes, Duméril, Duvernoy, Flourens. It was also under the impetus given by Cuvier's mind that the younger men were formed: H. Milne-Edwards, A. de Quatrefages, Alcide d'Orbigny, etc. Cuvier's continuers, in an effort to defend the master's doctrines, often show even more dogmatism and intolerance than he himself. Geoffroy Saint-Hilaire was to regret no longer having Cuvier in person as an opponent of his own views. Cuvier professed to adhere to positive facts. Geoffroy claimed the right to let himself be guided largely by hypotheses and ideas. Inspired by Cuvier's continuators, every theoretical idea fell under suspicion, and, for many tedious years, zoölogy was to be imprisoned within narrow views. This state of mind explains the continued resistance that Darwin's theory was to meet with a little later, although evolutionist ideas had first been expressed in France by Lamarck. Cuvier's school spurned them unanimously. On the other hand, it is significant to find among the adversaries of Darwinism men who had not had any direct contact with Cuvier, but who were among the zealots of Comte's philosophy: such was the case of the anatomist and histologist Ch. Robin, who had not, however, the preoccupation of religious orthodoxy to alienate him from transformism.

In the front rank of zoölogists of the post-Cuvierian period must be placed Henry Milne-Edwards (1800-1885), who showed much less exclusivism than Cuvier's pupils, and who often tried to reconcile the latter's views and those of Geoffroy Saint-Hilaire. Besides, H. Milne-Edwards left first-rate studies in the field of descriptive zoölogy, and from his long teaching experience came his *Leçons sur la Physiologie et l'Anatomie comparée des animaux*,[1] which is a living monument to the learning of his time. After setting out as a zoölogist, Armand de Quatrefages (1810-1892), turned toward anthropology where he did im-

---

[1] 14 volumes. Paris (Masson), 1857-1883.

portant work; he, too, made a great effort to be impartial in the controversies of that period. Isidore Geoffroy Saint-Hilaire (1805-1861), son of Étienne, was inspired by his father's ideas quite liberally. During this generation, one of the most personal and most sound works was that of Félix Dujardin (1801-1860), an excellent observer of inferior organisms, who was the first to have a clear idea of the structure of living matter and described it in 1839 under the name of *sarcode*, for which name was later substituted, for no reason at all, that of *protoplasm*. Dujardin left excellent studies on the various parts of the zoölogy of invertebrates, particularly on Infusoria, where with great perspicacity, he rose against Ehrenberg's erroneous views.

A little later, Henri de Lacaze-Duthiers (1822-1901), completed a great zoölogical work, both by means of his extended investigations in the comparative anatomy of invertebrates, where he was able with great acuteness to interpret aberrant types, and by means of the rôle of stimulator that he played by creating a school of numerous adherents during many long years. He was, besides, one of the pioneers of marine zoölogy and the most active and most successful promoter of the establishment of maritime stations, the founder of those of Roscoff and Banyuls. His pupils were in turn the masters of the science of zoölogy in France during the last third of the nineteenth century: Edmond Perrier (1844-1921), and Alfred Giard (1846-1908), who were the first to spread the transformist ideas in France. Giard was one of the most comprehensive biologists of his time, having an astonishingly vast knowledge of zoölogy and botany, as well as a keen understanding of ethology and general problems. He, in turn, was the leader of a school. Before being converted to transformism, Yves Delage (1856-1920), had shown himself to be an anatomist of the most expert kind and a keen observer. One of his most remarkable studies was the minute discovery and study of the altogether abnormal development of the *Sacculina*. He afterwards had a large share in the investigations of experimental biology at the beginning of the present century, and, after J. Loeb, particularly in those of experimental parthenogenesis. The zoölogists of the same period who would deserve to have their names recalled, would be numerous: Léon Dufour, H. Fabre, J. Pérez, and many other entomologists, Balbiani, Georges Pouchet, A. F. Marion, Camille Dareste, one of the founders of teratogenesis, L. Chabry, the creator of experimental embryology, Em. Maupas, whose investigations on Infusoria have marked an era, as have his studies on sexuality in divers groups of invertebrates; L. Ranvier, one of the great masters of modern histology.

Botany.

Botany, in all its branches, has had representatives of great merit. We must confine ourselves to mentioning some of them rapidly, such as the Tulasne brothers in mycology, G. Thuret (1817-1875), and Ed. Bornet (1828-1911), in algology, Mirbel (1776-1854), Ph. van Tieghem (1839-1914), in anatomy and in very diverse directions, Gaston Bonnier (1853-1922), in experimental anatomy, Dutrochet (1776-1847), the physiologist to whom we owe the discovery of osmosis, L. Guignard (1852-1928), one of the protagonists of vegetable cytology, Noel Bernard (1874-1911), who died prematurely, and who had already distinguished himself by splendid investigations on symbiosis, the first to succeed in cultivating orchids directly from the seed. And, if France has recently had only a limited share in the development of genetics, it is, however, French scholars to whom we owe certain investigations that are fundamental to this science: Alexis Jordan (1814-1897), to whom we owe the idea of elemental species, and Charles Naudin (1815-1899), who shares with Mendel the discovery of the essential laws of hybridization. We must add to their names that of Ph. André and L. Levêque de Vilmorin (1776-1862 and 1816-1860), whose experimental investigations on the wild carrot and on the improvement of sugar beets are also one of the basic documents of modern genetics. Since the eighteenth century, the Vilmorin family has successfully combined scientific research work and the commercial growing of seeds. It has thus obtained a documentation of the highest importance in the horticultural field.

Paleontology.

Both zoölogy and botany have an inseparable supplement in paleontology, which, especially at the beginning, found an extremely favorable field in France. Because of his admirable studies on fossil vertebrates in the vicinity of Paris, Cuvier is often considered the founder of paleontology. In any case, he did determine its development. The osteography undertaken by Blainville and, unfortunately, not completed, was a happy supplement to Cuvier's studies and has remained a precious source of information. Lamarck had produced, on invertebrates, a work similar to that of Cuvier. Alcide d'Orbigny (1802-1850), produced, in turn, an important work of classification and description. We can get an idea of the activity of the paleontological investigations in France through the statistics given by d'Archiac and evaluating at more than 5,800 the number of plates published in France on paleontology between 1823 and 1867. This movement has scarcely slackened. Paleontology, particularly that of vertebrates, has, during the course of the nineteenth century, given rise to a number of important works in France, such as those of Albert Gaudry (1827-1908), of Ed. Lartet (1801-1871), of P. Gervais (1816-1879), of H. Filhol (1843-1902), of Ch. Depéret (1854-1929). One of its chapters, and by no means the least interesting one, human paleontology, found some of its most eminent protagonists in France. It suffices to mention the names of M. Marcelin Boule. Prehistory is a science of French creation, through Boucher de Perthes (1788-1868), who had numerous successors: Lartet, de Mortillet, Cartailhac, Ed. Piette, etc., to make the most of the very fine and very numerous beds afforded by French soil.

Vegetable paleontology had as its founder Adolphe Brongniart (1801-1876), son of Alexandre Brongniart, Cuvier's associate, through the *Prodrome d'une Histoire des végétaux fossiles* published in 1828. We must mention among his most conspicuous successors the Marquis de Saporta (1823-1895), W. Schimper (1808-1880), C. Grand-Eury (1839-1917), C. Eg. Bertrand (1851-1917), B. Renault (1836-1903), R. Zeiller (1847-1915), O. Lignier (1855-1916), all of them men to whom we owe a good number of the most significant results attained at this time relative to plant fossils.

Geology.

Paleontology is related, no less closely than to zoölogy and botany, to geology. Since the beginning of the nineteenth century, this latter science has been developed in France with a zeal and success that have never diminished. In 1930, the *Société Géologique de France* celebrated its centenary, and the account of its hundred years' activity gave the measure of the part France has taken in the progress of geology. The center of geological research work has been the study of French soil and subsoil, a study which has been condensated in large-scale geological maps of the region: this collective work has led, on the way, to the asking and solving of a number of problems of general geology. Among the most eminent workers of this century of research, we may mention the names of Constant Prévost (1787-1856), of Elie de Beaumont (1798-1874), whose too dogmatic general doctrines reigned and weighed on all geology for a long time, Ed. Hebert (1812-1890), one of the principal founders of French stratigraphy, Ch. Lory (1823-1889), one of the pioneers of alpine geology, Jules Gosselet (1832-1916), Marcel Bertrand (1847-1907), E. Munier-Chalmas (1843-1903), Em. Haug (1861-1927), W. Kilian (1862-1925), G. Vasseur (1855-1915), P. Termier (1859-1930), Ch. Depéret (1854-1929), Charles Barrois, and many others should be associated with them. F. Fouqué (1828-1904), and A. Michel-Lévy (1844-1911), devoted themselves especially to the study of volcanoes and eruptive rocks and were the founders of petrography. M. L. Cayeux is now a master of the study of the structure and origin of sedimentary rocks.

Having had French territory as its main field at first, the activity of French geologists now extends more and more to France's colonial domain, to Algeria, Tunis, Morocco; North Africa has thus been the scene of the work of Pomel, Ficheur, Gentil, etc. The Sahara has been studied by Rolland, Flamant, Chudeau, E. F. Gautier; the geological exploration of Madagascar, Indo-China, Syria, and New Caledonia is in full swing. The whole of the results obtained in the geology of the entire world has in France been the subject of two great synthetic works in the form of treatises, and which have been authoritative: that of Albert de Lapparent and that of E. Haug. France certainly is one of the countries having contributed in a most important way to the progress of geology.

Physiology.

Observation, pure and simple, of phenomena and facts and their description was, for a long time exclusively the sole study procedure in the biological sciences; it is only since the beginning of the nineteenth century that the experimental method has penetrated them. A large part of these sciences, however, necessarily remain subject only to the descriptive method. It is in the study of the functions of organisms that experimentation occurs, and physiology was thus constituted. Functions often follow with evidence of the anatomical knowledge of organs; but the mere consideration of anatomy is generally insufficient and leads to more or less ingenious, but uncertain and sometimes false, interpretations. Numerous functions—we see it more and more—and especially the most important ones—do not come to light in the anatomical investigation. We can, moreover, state in retrospect that, as long as experimental physiology did not exist, a series of absolutely artificial systems, relative to the interpretation of the functioning of the organism, followed. Through his natural, gifted intuition, Descartes imagined that the entire working of the organism ought, in the last analysis, to come down to mechanics, that is to say, in our present-day language, to physico-chemical phenomena. But the manner in which he pictured this functioning to himself, taking his cue from anatomy, is entirely imaginative. Iatromechanism and iatrochemistry, which arose from the Cartesian doctrines, and which flourished in the seventeenth and eighteenth centuries, are no less imaginative. The idea of experimenting on living beings is, however, quite ancient. The Alexandrian school used it in olden times. Galen, in the second century, was the father of experimental physiology. But, aside from some forerunners, physiology is an altogether modern science, a science of the nineteenth century. In France, in the eighteenth century, there were several sporadic attempts at experimentation. Thus, in 1710, Poufour du Petit (1666-1741), proved by direct experiments the crossing over of the cerebromedullary fibers; by injuring the cerebral surface on one side, he caused paralysis of the side opposite to that of the lesion. In 1748, investigating the medulla oblongata, Lorry (1725-1786), discovered a definite point, at the base of the fourth ventricle, where a puncture suffices to cause death immediately. In this point was seen the seat of the soul, which had previously been located in various places. We know to-day that this point is merely a nervous center, essential to the coördination of respiratory movements. Réaumur, too, was a physiologist, and even more so. With his constant care for maintaining a contact with reality, and because of his training as a physicist, he was often led to methodical experimentation. Thus he performed the work of a real physiologist on the digestion of birds. But in the eighteenth century, the place of honor in this respect belongs to Lavoisier, who, through his investigations of respiration and animal heat, is the founder of general physiology and biological energetics. Thus we find only isolated attempts. It was in the nineteenth century that physiology was to emerge as a science in its own right, with a fixed doctrine, and it was in France, more than anywhere else, that this great stage of scientific development was reached.

A phantom arose against it, as we can well see to-day, that of vitalism, a doctrine according to which a special life force ruled all the phenomena of the organism and made of it a system radically different from the inorganic world. It was the universal belief, and that professed particularly by the Faculty of Medicine of Montpellier, then in all its glory, through François de Sauvages (1708-1767), Theophile de Bordeu (1722-1776), and Paul Barthez (1734-1805). Finally, at the very threshold of the nineteenth century, these vitalist ideas were asserted by a gifted mind, the founder of general anatomy and the precursor of histology, Xavier Bichat (1771-1802), whose life was too short really to enable him to give his full measure and free himself from the imprint of the doctrines on which he had been trained. Bichat had the very novel and fruitful idea of attributing all the properties of the organism to the elemental properties of tissues. But in the functioning of the latter, he retained the vital force; he was, then, content, as has been said, to decentralize it. According to him, tissues had two classes of separate properties: the former vital and unstable "sans cesse variables dans leur intensité, leur énergie, leurs modifications . . ." the latter, "propres au monde physique, fixes, invariables, sources de phénomènes uniformes, susceptibles d'être soumises au calcul . . ." "On calcule," did he add, "le retour d'une comète, les résistances d'un fluide parcourant un canal inerte, la vitesse d'un projectile, etc. . . . mais calculer avec Borelli, la face d'un muscle, avec Keil la vitesse du sang, avec Jurine, Lavoisier la quantité d'air entrant dans le poumon, c'est bâtir sur le sable mouvant un édifice solide par lui-même, qui tombe bientôt, faute d'une base assurée."[1] Life, according

---

[1] The former vital and unstable "incessantly variable in their intensity, their energy, their modifications . . ." the latter, "peculiar to the physical world, fixed, invariable, the sources of uniform phenomena, capable of being subjected to computation. . . ." "We forecast," did he add, "the return of a comet, the resistance met by a fluid traveling through an inert channel, the speed of a projectile, etc. . . . but to compute with Borelli the force of a muscle, with Keil the speed of the blood, with Jurine and Lavoisier the quantity of air entering the lung, that means building on moving sand a structure firm in itself, which soon falls for want of a secure foundation."

to Bichat, is merely the conflict between vital and physical forces. According to his famous definition, it is the whole mass of functions that resist death. Death is the ultimate triumph of the physical over the vital forces. Such language, immediately following Lavoisier's discoveries, is astonishing in a man of Bichat's worth, and it is probable that, if he had lived much longer, his ideas would have been modified a good deal. But this vitalism will still persist a long time. Johannes Muller, for example, one of the men of whom German science is most proud, still formally professed it in 1835. It is, then, necessary to recall this state of mind in order to evaluate all the importance of the revolution that was effected during the course of the first half of the nineteenth century, destroyed this phantom, and was above all the work of the French physiologists.

Legallois, a physician of Bichat's generation, but who lived a little longer (1770-1814), and who, in the midst of great material difficulties, experimented on animals in the rare moments of leisure that his profession left him, already determined the rôle of the so-called vital spot of Lorry; he even had an intuition of the importance artificial respiration would have for experimentation. Despite the encouragement given by Laplace, he was, unfortunately, obliged to give up experimentation. But a new champion appeared almost immediately in the person of François Magendie (1783-1855). Independent of spirit, tempered by the difficulties his youth had encountered, he attacked Bichat's vitalist ideas as early as 1809. Resolved never, in his conclusions, to exceed the limits furnished by experience, he experimented and sought the explanation of all the phenomena presented by the organism, in physical and chemical laws. Pushing further his skepticism respecting theories, he refused to admit anything but facts. "Chacun," he said familiarly, according to what Claude Bernard reports, "se compare dans sa sphère à quelque chose de grandiose, à Archimède, à Michel-Ange, à Newton, à Galilée, à Descartes, etc. . . . Louis XIV se comparait au soleil. Quant à moi, je suis beaucoup plus humble. Je me compare à un chiffonnier. Avec mon crochet à la main et ma hotte sur le dos, je parcours le domaine de la science et je ramasse ce que je trouve."[1] Animated with this spirit, Magendie was the true founder of the experimental method in physiology. He opened, at first, in Paris, a private laboratory, to which many foreigners came to develop themselves, several of whom have since become masters, such as the great German physiologist Ludwig. We notice here, once again, the clearsightedness of Laplace, who discerned, without any delay, the worth of Magendie, granted him his patronage, and facilitated his entrance to the *Académie des Sciences,* to which Magendie was elected in 1821. Magendie's work would be interesting to study in detail. I recall only that to him belongs, as Claude Bernard has shown,[2] the rigorous demonstration, made in 1827, of the functions of the roots of the spinal nerves, which functions the English physician

---

[1] "Each one," he said familiarly, according to what Claude Bernard reports, "compares himself in his sphere, to something grandiose, to Archimedes, to Michael-Angelo, to Newton, to Galileo, to Descartes. . . . Louis XIV compared himself to the sun. As for me, I am much more humble. I compare myself to a ragpicker. With my hook in my hand and my basket on my back, I travel over the domain of science and I gather up what I find."

[2] The rights of Magendie to this discovery have just recently been proven by the English physiologist Aug. Waller (Meeting of the British Association, Portsmouth, 1911). Waller has shed light on the very significant fact that Bell re-edited textually, in 1824, certain works dating from 1821 and 1822, altered the text without informing the reader, so as to make it agree with the discovery of Magendie in 1822. That also is visible on one of the plates (plate III), and in his explanation, in which appeared for the first time, in Bell's writings, the distinction between motor nerves and sensory nerves. Quoted from the analysis of Waller's work by L. Frédéricq. *Revue générale des Sciences.* Paris. T. 23, 1912, p. 462.

Ch. Bell apprehended, but in a rather inaccurate way, as early as 1811. Magendie rediscovered the facts without knowing the works of Bell; the series of Magendie's publication on this subject, his doubts, his restrictions, are luminous proof of the precision of his method. His pitiless demand of rigorous proof led him to complete his initial discovery by that of recurrent sensibility which explained the apparent contradictions among certain of his observations.

Another remarkably clever and imaginative experimentalist was Flourens (1794-1867), who made some important discoveries as a result of bold and well conducted experiments, about the physiology of the brain, about that of the semicircular canals, and about many other problems. But Flourens did not possess Magendie's assuredness of doctrine. He did not know how to discipline himself, as Magendie did, and he sometimes allowed himself to be carried along to an adventurous dogmatism, which led him to make himself the irreconcilable defender of theories destined to be destroyed.

Claude Bernard.

The great figure of French physiology, and one can say of all physiology of the nineteenth century, was Claude Bernard (1813-1878). Formed by the rigorously positive discipline of Magendie, he was, according to the words of J. B. Dumas, Physiology itself. In a few years, he renewed or propounded the greatest problems. First of all, he renewed the physiology of the digestive juices. He discovered and thoroughly investigated the part played by the pancreas in the digestion of fats. Then, by the study of the digestion of sugars, he arrived at the capital discovery of the glycogeneous function of the liver, unsuspected up to him, and which dominates the entire inmost nutrition of the organism, regulates the constancy of glucose in the blood, assures the provision of glucose to the muscles, that is, the reserve of energy which is necessary to them. He thus fully elucidated the problem of animal heat, the foundation of which had been laid by Lavoisier and Laplace. The impor-

Claude Bernard

tance and the novelty of these discoveries are measured according to the discussions which they have provoked. Claude Bernard refuted all the objections which were raised. The glyco-genesis of the liver is the first example of an internal secretion, a notion of which we see to-day the full expansion and the capital value. At the same time, the idea of a constant interior "milieu" is evidenced, in which live the anatomical elements, which are really free, in superior animals, from all the physical and chemical fluctuations of surrounding conditions. No less remarkable are his discoveries in the physiology of the nervous system. He showed its all-important action on the circulation of the blood, the great sympathetic system regulating the diameter of the vessels, contracting or dilating them, and in this way adjusting the flow of blood in the organs. Then it was the varied study of a long series of poisons—strychnine, curare, etc., from which he drew marvelous means for studying the physiology of the nerve and of the muscle; it was also the study of carbon monoxide by means of which he made clear the properties of hæmoglobin.

Above all these discoveries, any one of which would have sufficed to make him illustrious, he established, on indestructible foundations, against various supporters of vitalistic ideas, the doctrine of *experimental determinism* in physiology. There is no caprice in the functioning of the organism. Its laws are precise and rigorous, like those of the physical world and the phenomena of life obey physical and chemical laws only, but in the conditions under which the latter are operating are much more complex. Biological experimentation is the difficult art of knowing how to render constant all the conditions of the experiment except one and to know how to make that very one vary whose effects one wants to study. That realized, the organism responds in a strictly determined fashion. Not only did Claude Bernard practice this art with an incomparable mastery, but he fixed these principles in a book which, in the history of biological sciences, deserves a place comparable to Descartes's *Discours de la Méthode, l'Introduction à l'étude de la Médecine expérimentale* (1865). He determined what the mentality of the experimentalist should be, free of all spirit of system, always placing himself under the control of methodical doubt, without skepticism, and surrounding himself ceaselessly with definite precautions, the most fundamental of which is the proof by the converse experiment. It is sufficient to recall what was said of Bichat, in order to measure the ground covered. Physiology, in the hands of Claude Bernard, has become a science as precise and exact as physics or chemistry, but requiring much more delicate handling. More than half a century away, *l'Introduction à la Médecine expérimentale* has but increased in importance.

The physiology of Claude Bernard in its enormous

breadth remains essentially positive. It studies and determines the conditions of the interrelation of phenomena, without burdening itself with metaphysical preoccupations on the essence of things. Like so many other manifestations of French science in the nineteenth century, it is closely connected with the thought of Auguste Comte, who, in 1836, before the appearance of Claude Bernard and perhaps under the influence of Magendie, had risen against the vitalistic conceptions of Bichat and had, in his *Cours de Philosophie positive*[1] declared with a magnificent clear-sightedness that the moment had come when physiology was to be liberated not only from metaphysics but also from medicine. That is what Claude Bernard effectively realized, and as one of his pupils and successors, A. Dastre, very expressively put it on the occasion of the celebration of his centenary in 1913: "Il a chassé de la physiologie les fantômes qui l'encombraient. Elle était l'humble servante de la médecine. Il en a fait une science indépendante, ayant ses méthodes et son but. Il a accompli une révolution dont les nouvelles générations ne se doutent pas, parce que les résultats en sont si bien acquis qu'ils font en quelque sorte partie de notre mentalité et que, selon le mot de Montaigne, l'habitude en ôte l'étrangeté."[2]

This immense work was accomplished under the worst material conditions, practically without resources and in miserable and unhealthful quarters. In reference to this, one could not read too often the magisterial report written by Claude Bernard for the occa-

---

[1] Tenth lesson.

[2] "He excluded from physiology the phantoms which encumbered it. It used to be the humble servant of medicine. He made an independent science out of it, having its own methods and its purpose. He accomplished a revolution of which the new generations have no idea, because the results of it are so well acquired that, in a way, they form part of our mentality, and to such a point that, according to the saying of Montaigne, habit removes strangeness."

sion of the Universal Exposition of 1867, *Sur les Progrès de la Physiologie en France*. He sadly compares the misery in the midst of which the French scholars worked and the well-equipped and well-provided-for laboratories already at the disposal of their German colleagues and, in particular, the physiologists. During the nineteenth century, in which experimental sciences, which up to then had hardly been but a very clever pastime for original minds, fully revealed their limitless fruitfulness for increasing the power of man and bettering the human condition, one of the weaknesses of French science, which we must not conceal, was the insufficiency of material resources put at the disposal of its great creative minds. If some among them succeeded, nevertheless, in accomplishing their work, one may think that it could have been done more easily and more completely with the aid of better appropriated means, and it cannot be denied that this indigency must have stopped more than one vocation.

If Claude Bernard was incontestably the greatest French physiologist of the nineteenth century, he was far from being an isolated one. First of all, he left pupils who have brilliantly continued his work: Paul Bert (1833-1886), to whom we owe especially very important works on respiration, but whose activity was quite rapidly diverted toward politics; N. Gréhant (1838-1910), who applied himself above all to the gases of the blood; A. Dastre (1844-1917) and J. P. Morat (1846-1920), who have especially continued the master's work on vasomotor nerves; A. d'Arsonval, who turned particularly toward applications of electricity to physiology, etc. And, outside of his school, we must not disregard the great value of a certain number of his contemporaries: J. Marey (1830-1904) was an experimentalist with marvelous ingenuity, who also learned how to make up for the dearth of material means, and to whom we owe, if not the invention, at least the considerable extension of the graphical method in physiology. He knew how to imagine the proper devices for its use in the most varied cases. He was, above all, the creator of the physiology of motion, in realizing for each case an indelible and authentic concrete representation. Marey was thus the forerunner of cinematography, which he came very near to realizing definitely, before the brothers A. and L. Lumière did it (1895). A place of honor among physiologists of the same period must be reserved for Aug. Chauveau (1827-1917), whose long career, begun with Marey, in the study of the motions of the heart, was particularly devoted to the general energetics of the organism. He also had the honor of having been a forerunner in the study of viruses, which was illustrated by Pasteur. Finally, Brown-Séquard (1819-1894), whose double name recalls his English father and his French mother and whose life, for a long time, was a succession of alternate attempts of scientific establishment in France and in the United States. As a matter of fact, he was a professor at the University of Virginia in 1852; then after a period of time spent in France, he taught at Harvard University in 1863. In 1878, he succeeded Claude Bernard at the *Collège de France*. The work of Brown-Séquard is varied and interesting. What we shall particularly retain from it is that, at the end of the century, it gave impetus to the study of internal secretions, thus developing in an extremely rich way, an order of facts the foundation of which was laid by Claude Bernard, through the discovery and the study of the glycogenic function of the liver.

The Work of Pasteur.

If Auguste Comte had felt the necessity of freeing physiology from metaphysics and from medicine,

Claude Bernard, from the beginning of his career, had called for the moment when medicine would rest on a scientific basis and would abandon empiricism. This revolution was to be accomplished under his eyes. It was reserved for Louis Pasteur, whose work counts among the most considerable, the most fruitful and the most beneficent that science has yet realized. It is so well known and so popular that, pressed as I am by time, I shall be permitted to limit myself to merely recalling its main trends, throwing the light on the logical character of its sequence and its essential characteristics. And firstly, as quite frequently happens to those who are destined to reform a science, Pasteur was an outsider, a stranger to medicine, just as Lavoisier was not a professional in chemistry. Superior minds are more receptive to ideas, if they have not been enveloped at the beginning by reigning ideas and by explanations presented as dogmas. Such was the case particularly for medicine, which from Hippocrates up to the end of the nineteenth century consisted of a series of systems which succeeded each other and each one of which was as complete and as artificial as the other.

The Hemihedrism of Tartrates.

Having entered the *École Normale* in 1843, without having been up to then, and without being a brilliant pupil at the *École Normale,* Pasteur had turned there toward chemistry and his first investigations were the occasion of an important discovery. It was in the field of crystallography and of physics. Tartrates gave some solutions endowed with rotatory power for some, neutral for others. These anomalies had checked the wisdom of men like Mitscherlitch. Pasteur, in studying the crystals with the microscope, discovered that they were dissymmetrical, presenting

certain facets, turned in some to the right, in others
to the left, the first being the images of the second
in a mirror.   He thus discovered the hemihedrism of
tartrates.   He then had the idea of sorting, under the
microscope, the two kinds of crystals; then making
solutions of each one of these categories, he saw that
the direction of the rotary power corresponded to that
of the hemihedrism, and that by combining equal
weights of the two kinds of crystals one obtained a
neutral solution.   Rotatory power appeared then to
be directly connected with the dissymmetry of the
crystals, that is, with the dissymmetry of the molecules, the first and daring transposition of a problem
from the microscopic scale to the molecular scale.
Thirty years later, these results served as the starting

Louis Pasteur

point for Le Bel and for van t'Hoff to conceive the tetrahedral structure of the atom of carbon and to elaborate stereochemistry, a magnificent extension of the atomic theory.   It was along this unexpected path that Pasteur was to be impelled toward biology.   The fact soon impressed itself on his mind that molecular dissymmetry was the feature of substances elaborated by living beings, whereas it did not exist in substances secured outside of life, in laboratories.[1]   The living cell appeared to him as a field of dissymmetrical forces, realizing under the influence of cosmic forces, equally dissymmetrical (the direction of the rotation of the earth by itself introduces a dissymmetry into a series of phenomena) transformations which are themselves dissymmetrical.   Some molds, vegetating at the expense of neutral solutions of tartrates, destroy by their nutrition one of the categories of molecules to the exclusion of the other, and finally leave an active solution either to the right or left.   Nutrition is then related to molecular dissymmetry.   There was here a field of problems of considerable importance which had inflamed the imagination of Pasteur, but from which he was turned away, not without regret.   It is probably a source from which, in the future, a penetrating intelligence will extract important discoveries.

Fermentations.

Pasteur, after a few years' sojourn at Strasbourg, where he had continued his work on tartrates, had been appointed to the faculty of Lille, and there sev-

---

[1] There is a restriction to be made to-day in this respect, following the works of Jungfleisch, who showed that the various tartrates, heated in the presence of water are transformed into one another; he also succeeded in preparing racemic acid (inactive) by synthesis, from ethylene, transformed at first into succinic acid, that is outside of all intervention of living beings.   As one can, from there, pass to active acids by heat, one sees that life is not indispensable to the acquisition of a dissymmetrical product.

eral distillers soon took advantage of his knowledge in connection with irregularities occurring in the fermentation of industrial alcohol. This industry had remained until then in a total state of empiricism. By his intervention Pasteur was about to make some capital discoveries. The mechanism of fermentation was as yet unknown. We to-day can appreciate the merit of predecessors who had perceived therein little globules in the process of multiplication; in the seventeenth century Leuwenhoek, the father of microscopy; in the nineteenth, in 1831, in France, the chemist Cagnard de Latour, and, in Germany, Schwann. Cagnard de Latour had even noticed their multiplication through germination and had considered yeast as an organism, whose vegetation was probably to bring on the setting free of carbonic acid and the formation of alcohol. Schwann had propounded similar ideas. But these observations had remained without consequences, and two theories—one of Berzélius, the other of Liebig —were struggling for preference; both were obscure, purely verbal, and impregnated with vitalism. From the start, Pasteur realized that the difficulties which he was called upon to remedy were correlatives of the development, at the surface of the wort, of a pellicle which the microscope showed to be made up of an accumulation of little grains constituting an organism, in the process of multiplication; and besides, of the production of lactic acid. Instructed by his studies upon tartrates and the growing of molds upon them, he sowed this film into a sugared solution and easily obtained again, in this way, the film and lactic acid. This acid was the product of the organism of the film nourishing itself from the sugar of the solution; it was the residue of this nutrition. The practical question which had been proposed for him was solved. In order to obtain a good fermentation, it was necessary to escape the presence of this parasitic organism. The fermentation itself appeared to him as related to the nutrition of an organism. In the course of his experiments he had secured lactic fermentation at the expense of a special organism: the lactic ferment. Alcoholic fermentation was the result of the life and multiplication of globules of yeast which he found in the wort, according to Cagnard de Latour, and not a product of death or putrefaction, or of mysterious motions, as Berzélius and Liebig would have it. The memoir of Pasteur on lactic fermentation, presented to the Society of Sciences, Agriculture and Arts of Lille, in its fifteen concise pages, contained the germ of all the work of Pasteur that followed, and marks an important date in the history of science. In effect, he had lifted a corner of the veil of mystery which had balked Lavoisier,[1] the restitution to the mineral kingdom of substances found in the bodies of plants and animals; thus the work of Pasteur appears as a direct continuation of that of Lavoisier. Pasteur himself thought so, as shown by a remarkable writing of his recently published.[2]

The signal discovery resulting from the work of Pasteur upon alcoholic fermentation is the exact relation established between the development and multiplication of a given organism, and the production of a determined chemical compound. Pasteur had evidently been guided by the results of his previous studies with tartrates. In both cases, the same instrument had been the decisive means of discovery and verification, i. e. the microscope. Pasteur had introduced it into the stock of the chemist's tools, as had Lavoisier the balance. Fermentation is now an exact process resulting from the action of an organism, the

---

[1] Cf. Second Lecture, p. 55.

[2] Pasteur—*Note à M. le Ministre de l'Instruction publique* (avril, 1862), *publiée à l'occasion de son centenaire* (*Revue de Paris,* 15 juillet, 1923, *et Bulletin de l'Institut Pasteur,* T. 21, 31 juillet, 1923).

ferment, upon a nutritive substance appropriated and fermentable. There is obtained without failure the product of the fermentation under favorable conditions, by bringing together ferment and fermentable substance in the pure state. In nature, numerous ferments vegetate side by side in complex environments. There is exercised among them a competition which eliminates one to the profit of the others, as was being proclaimed at the very moment for the totality of organisms, by Charles Darwin.

The concept is of a general value in the field of fermentation, and Pasteur, in a few years, applied it to a series of successive cases, extracting in the process some new discoveries of major importance. It is in this manner that he studied the production of butyric acid; he discovered therein a fermentation with its specific ferment, the butyric bacillus. This ferment is mobile, but, above all, it has the unforeseen peculiarity of not being able to exist except protected from the air; it was the first example of an anærobic organism; and in this manner was born the notion, at once most significant and paradoxical, of *life without air,* i. e. *anærobiosis,* where, in reality, respiration by the help of the oxygen of the air—this fundamental function of life—is replaced by a borrowing of oxygen from the combined state. Soon, the alcoholic function itself was placed inside the limits of anærobiosis. Beer yeast can exist, in effect, as Pasteur established, either in the ordinary manner as an ærobia and it then burns up sugar completely, without leaving it in the state of intermediate oxidation which is alcohol, but transforming it into carbonic acid and water; or it may live in the manner of an anærobia, sheltered from air, and it is then that it produces alcohol by an incomplete combustion. Yeast is an optional anærobia, forming the transitions of ærobia to pure anærobia, like the butyric ferment.

Fermentation is life without air. Furthermore, the pure anærobia furnished Pasteur with the explanation of another great natural phenomenon, putrefaction. This time he completed the explanation of the restoration to the mineral state of organic substances. Thus, the cycle of the transformations of matter through life is closed; thus, the mystery of which Lavoisier had had the intuition and had perceived the principal trends is definitely solved.

The practical consequences of the study of fermentation are multiplying and transforming a series of industries. The manufacture of vinegars, until then empirical, was brought back to the vegetation of a specific ferment, the *Mycoderma aceti,* at the expense of wine, with production of acetic acid by partial oxidation of the alcohol; so, the problem of regularity in manufacturing is reduced to securing a pure ferment. The manufacture of beer was also transformed and regularized by the use of pure yeasts, of a fixed type, giving a determined speed to the fermentation. All the alterations and contaminations of wines, the souring of Bordeaux wines, the bitterness of Burgundy wines and the fattiness of Champagne wines, were explained by the growth of definite ferments; the practical means which Pasteur found to avoid them was to heat the wines to 55° C., thus destroying the harmful ferments. It is the process currently used in industry which has received the names of *pasteurization* and which is applied each day on a large scale to the preservation and transportation of perishable liquids such as milk.

Spontaneous Generation.

In the years around 1860, an age-old controversy was revived, that about spontaneous generation. The *Académie des Sciences* raised the question for a contest and certain members of the *Académie* urged Pasteur to attack it, while others feared to see him engaged in the study of a problem considered by them insoluble. Pasteur conducted, with some simple and novel apparatus, a series of experiments of an impeccable precision and logic, which proved the worthlessness of the alleged facts in favor of spontaneous generation. No organisms can be found, except where germs have been introduced or were existent and not previously destroyed. Several years later, the quarrel was revived on the basis of new facts pointed out by a new champion, the Englishman, Charlton Bastian. Pasteur again emerged victorious from this new discussion, explaining the assumed spontaneous generations of Bastian by a new discovery, that of germ resistance to a temperature above 100°. These discussions have not only settled in a definite manner the question of spontaneous generation, they have had theoretical and practical consequences of major importance; from a theoretical point of view, there is no development of organisms without a pre-existing germ; from a practical point of view, it is from these investigations and discussions that the whole technique of bacteriology arose, of which the essential base is the sterilization of the cultural media. A complete stock of implements, now employed in laboratories the world over, has been created. In particular, the steaming pan *(autoclave)* was conceived and constructed by Chamberland, a collaborator of Pasteur, at the time of the discussion with Bastian, in order to obtain the sterilization of the cultural media, at a temperature above 100°.

Each step in the career of Pasteur was based on the preceding one and combined the solution of a great theoretical problem with that of a practical question. We find this double character in the works on tartrates, on fermentations. on spontaneous generations.

We find it again in another instance. In 1865, a disease of the silkworm, the *pebrine,* against which all efforts were useless, was ruining French sericiculture, the wealth of the southeastern part of the country. Pasteur was urged to undertake to discover a remedy. But he was neither a zoölogist nor a sericiculturist. He had never handled silkworms; he had no notion of their breeding. Was it not a gamble to risk himself in interfering in this domain? Nevertheless, if he neither sought to solve nor solved the problem of zoölogical order which was at stake and which has since been clarified by degrees,[1] he very rapidly solved, in a personal and purely experimental way, the practical problem of making safe the breeding of the butterfly. His sagacity led him to discern an authentic indication of the disease, the presence of corpuscles which had already been observed, and which we have since learned to be the spores of the infecting parasite, microsporidia; he noticed that these corpuscles transmit themselves by the egg, and that if care is taken to pick for reproduction only grains (that is, eggs) free from these corpuscles, the breeding of the following generation will reach its term without damage, even if the worms are contaminated by the parasite in the course of their development. Pasteur thus enabled sericiculture to subsist; but, more important still, his investigations on the silkworm were of tremendous concern for the future, as being the first contact of Pasteur with pathogenic germs. For this disease caused to the silkworm by the corpuscles, the pebrine, was not the only one noticeable. Pasteur had to deal with others, such as the *flacherie,* which he was able to distinguish and trace to a multiplication of bacteria. These investigations, thus, led him into the

---

[1] The *pebrine* is caused by the multiplication into the tissues of the worm of a parasitic Protozoon, belonging to the class of the Sporozoa, a Microsporidia, the *Nosema bombycis.*

domain of pathology, where he was going to accomplish the enormous work which filled the second part of his career and made his name so popular, a work which has produced a revolution in medicine equivalent to that of Lavoisier in the domain of chemistry.

Infectious Germs (Viruses) and Contagious Diseases.

The germ theory resulting from his work on spontaneous generation and fermentation was beginning to make an impression on reflective minds. The Scotch surgeon Lister, as early as 1865, perceived the relationship that might possibly exist between germs diffused everywhere and the dreadful purulent complications of surgical wounds. He endeavored to protect himself from these with dressings preventing the growth of germs and inaugurated in this way the practice of antiseptics. In 1862, a French naturalist and physician, Davaine, enlightened by Pasteur's work on ferments, discovered, in the blood of animals affected with anthrax, a stringy germ which pullulated there and which he called *bacteridium*. It was the first example of a pathogenous bacterium. Toward 1870, Pasteur directed his activities to the hospital, suspecting that the purulence of wounds and puerperal complications were due to germs introduced into the organism, and that contagious diseases were the result of the transmission of germs. In a few years, this idea produced a series of decisive discoveries for each of which I cannot stop here.[1] Each contagious morbid entity becomes the pullulation in the organism of a

---

[1] One may find a detailed account of Pasteur's work most accurately recorded in the book of his son-in-law, R. Vallery-Radot: *La Vie de Pasteur* (Paris, Hachette), as well as in Duclaux: *Pasteur, Histoire d'un esprit* (Paris, Masson). The *Œuvres complètes de Pasteur*, are in course of publication, since 1923, under the care of his grandson, Dr. Pasteur Vallery-Radot. They comprise 8 volumes, of which 6 have already appeared. (Paris, Masson.)

specific germ just as all fermentation is the multiplication of a specific ferment in an appropriate nutritive medium. That is the great truth that asserted itself in the last half century.

Viruses and Vaccines.

But Pasteur was not to content himself with disclosing the origin of the disease; he was going to find, in handling the germ that causes it, the specific preventive remedy, and occasionally the curative one. After disclosing the virus, he made a vaccine from it. A disease played a leading rôle in this respect; it was anthrax, which man contracts sometimes, but which destroys cattle especially, thus occasioning enormous agricultural losses. In 1876, with his collaborators, Joubert, Chamberland, and Roux, he attacked the problem of anthrax. He isolated and cultivated on broth its infectious germ, the *bacteridium,* seen by Davaine some years before; by means of crucial experiments, he proved irrefutably that it was the one and only cause of the disease, refuting the different objections which, although based on experiments that were insufficiently exact, did not, at the time, lack force. In the course of his investigations, an unexpected circumstance was to be for him a ray of light. He had been led simultaneously to study another disease, hen cholera; and chance, by which penetrating minds can profit, furnished him with a disturbing fact. On his return from his vacation, he discovered that the germ of this disease, left alone in the cultural broth for two months, no longer killed the hens after inoculation. Intuitively, Pasteur realized that an attenuation of the virus had taken place as a result of the cultural conditions. By appropriate cultural modifications a virus was transformed into a vaccine, for the animals that had received an inoculation of attenuated virus resisted afterwards the inoculation of the virus ordinarily able to kill. The antivariolar vaccination by means of vaccine, that marvelous discovery which Jenner had made in the eighteenth century, and which had ever since remained an isolated fact, thereafter was no longer uncorrelated. Pasteur immediately applied the method to anthrax. He succeeded in attenuating the virulence of the anthrax *bacteridium* by heat, cultivating it at 42° or 43° C. That is how a vaccine is obtained. The result, however, was received with skepticism by certain people. In order to convince the skeptics, peasants and veterinaries, Pasteur accepted the famous Pouilly le Fort risky public experiment. Two flocks of sheep, consisting of twenty-five each, were inoculated with a deadly dose of anthrax virus: one flock was unvaccinated, the other had previously been vaccinated with virus that had been attenuated by heat. The sheep of the first flock were supposed to be dead by a certain date, for which a public meeting was called. On the appointed day, all the unvaccinated sheep, in fact, were dead, while the twenty-five vaccinated ones were in good health. Thence proceeds the universal practice of the antianthrax vaccination, which has prevented immense material losses. It was also the starting point of a new science, immunology.

Hydrophobia.

I have no time to stop for Pasteur's simultaneous investigations of other diseases. I must confine myself to calling to mind the fact that his glorious career found its completion in his cure for hydrophobia, which spread out his popularity. It was, after all, a marvel of sagacity and ingenuity, but one which does not attain the logical beauty of his previous discoveries, because there is involved in it a part of empiricism which, moreover, has not been reduced although it is almost fifty years since the practical problem was solved.

The weighty consequences of Pasteur's work on germs, from fermentation to infectious diseases, are so evident to-day that it is difficult to imagine the opposition with which these discoveries met in spite of the clarity and force of the proofs produced, and that, in spheres which should have been the first to understand and welcome them. This opposition, moreover, did not emanate from mediocre minds. As regards fermentation, Pasteur clashed with Liebig; as far as spontaneous generation is concerned, he had to struggle tenaciously with men who possessed, strange to say, a scientific education. It is this opposition, however, which happily culminated in bringing about a good part, at least, of the improvements in technique. But it is his work on infectious diseases, especially, that provoked violent opposition on the part of physicians. The Academy of Medicine of Paris was, in the years during which the work of Pasteur was being performed, the theater of veritable battles, and it took long years and the disappearance of a generation brought up in prepastorian ideas, before the medical world was completely won over to the side of the new doctrines, founded, as they were, on experimental evidence. The opposition did not come merely from physicians and Paris. Such eminent scientists as Virchow, Helmholtz, and Du Bois Reymond in Germany fought against Pasteur's ideas, claiming that they were too much tainted with vitalism. They wished for a mechanism of a purely physico-chemical order to explain infectious diseases.

It is the lot of all truly innovating discoveries, because of the prejudices and idols they overturn, to meet with obstacles, by reason of their very fecundity.

The circulation of the blood raised a long and violent opposition in the seventeenth century; Newton's law of gravitation in the eighteenth century needed several decades before it could finally impose itself; Jenner's vaccination, in spite of its evident benefits, was also received with difficulty; Lavoisier's revolution in chemistry had as tenacious and, sometimes, irreducible adversaries the best chemists of the time; Fresnel's theory experienced prolonged opposition. It was the same with the atomic theory in chemistry, and here again, because of men who should have been the first to be convinced. All these controversies have a common aspect proceeding from their psychological side, and it is amusing to see springing up in them, at different times and in different subjects, the same expressions and the same comical element. It is thus that, seeming to copy the invectives of Riolan against Pecquet and of Macquer against Lavoisier, Peter, one of the physicians of the *Académie* who distinguished himself most woefully by the ardor and tenacity of his struggle against pastorian ideas, still wrote in 1883, that is to say, at the time when the proofs had accumulated: "Dans cette lutte que j'ai entreprise, l'affaire actuelle n'est qu'un engagement d'avant-garde, mais, si j'en crois les renforts qui m'arrivent, la mêlée pourrait bien devenir générale et la victoire, je l'espère restera aux gros bataillons, c'est-à-dire à la *vieille médecine.*"[1] In the case of Pasteur, this opposition is all the more significant because, all in all, it was not a question of theories grounded upon speculations more or less difficult to verify, but rather of tangible data. Pasteur never let himself be carried away by theory.

---

[1] "In this struggle I have undertaken, the present battle is only a vanguard engagement, but I believe, judging from the reinforcements coming to my aid, the fight might well become general and victory, I hope, will remain with the large battalions, that is to say, with the *old medicine.*"

He always knew how to discipline his very lively imagination, which suggested his discoveries to him intuitively, by a rigorous experimental method and a lasting concern for proof and fact. It will be noticed that he never cared about the essence of things, but only about the conditions under which facts could be linked with one another. He did not study the silkworm disease in itself, but interested himself only in the conditions of its transmission. And in the question of spontaneous generation it was not the question of principle that he considered, but the explanation of the positive facts attributed to it. This work, like most of the great French scientific achievements of the nineteenth century, bears then, the stamp of the positive method.

The pastorian revolution is beginning to sink into the past; the ideas or the results which appeared to the advocate of the old school of medicine as intolerable subjects of scandal are, to-day, of the order of banalities and commonplace topics. The work of Pasteur, nevertheless, is an advance which has not been paralleled since the beginning of the Hippocratic era, that is to say, for a period of twenty-five centuries. It marks the advent of a scientific school of medicine, that dream for the realization of which Claude Bernard was wishing. Even outside the field of medicine the work of Pasteur revealed to science a new world, that of those infinitely small beings, the *microbes,* a world which, if it contains redoubtable pathogenous agents, relatively easy to disclose, moreover, because of their very nature, is far from being limited to them. These infinitely small beings of life, which, until Pasteur, had been entirely unknown, play in nature an enormous part, a more effectual one even that the large organisms, and until Pasteur totally unsuspected. They are the great transformers of matter, those which restore to the mineral world organic substance, working on this collective work, each according to its own specific mode. It is a new domestication of these forces that flows from the work of Pasteur, a domestication he himself inaugurated for a series of industries which he drew away from empiricism. Many other applications of the same order have sprung or will spring from it. To cite only a few examples, the study of the bacteriology of the soil has changed the fundamental data of agriculture by regulating the exact conditions of the nutrition of plants. Already Boussingnault had had the intuition that there was in the soil an automatic mechanism by which it became rich in assimilable nitrogen. The nitrifying and denitrifying bacteria are the effective agents of this mechanism. Schloesing and Muntz, inspired by Pasteur's ideas, furnished, as early as 1876, an indirect proof of the fact that nitrification of the soil is a process of fermentation, by arresting it with the action of an anæsthetic. One of Pasteur's disciples, Winogradski, has since furnished the direct proof by isolating and cultivating the nitrifying microbes. Another of Pasteur's pupils, Raulin, made in 1870 a study of nutrition which has remained a fundamental and classic one, and in which he realized, in a synthetic manner, the optimum chemical medium for the development of a mold, the *Aspergillus niger* and the equivalent for a higher plant, maize, has been recently realized by another member of the pastorian school, M. A. Mazé. The botanist Van Tieghem, who belongs to the oldest generations formed by Pasteur, has introduced into mycologic cryptogamia the technique of pure culture which is now the basis of all research. Em. Duclaux (1840-1904), a lucid and critical mind, has carried these principles into the domain of biological chemistry. A number of other collaborators have made contributions to the master's work, supplementing and developing it in theory or application: the chemists

Gernez and Gayon, the physicist Joubert; Chamberland, the inventor of the steaming pan and of the filter arresting microbes; Thuillier, who died in Alexandria in 1883, while studying the cholera epidemic there. M. Émile Roux, to-day the only survivor of Pasteur's direct collaborators,[1] has magnificently prolonged the work of the master by discovering the first microbian toxins, thus widening tremendously the paths of immunology.

Pasteur himself, like all the men of his generation, executed his fundamental research work in wretched laboratories. He had at least the satisfaction of ascertaining that the echo of his discoveries would assure for his successors laboratories corresponding to the needs of the sciences he had created. In the institute which bears his name, and which has since opened branches in the French colonies, and even in several foreign countries, his tradition remains fecund. But to-day, the disciples of Pasteur are no longer limited to those who worked under his direction and those who people the laboratories of the Pasteur Institutes. They are all the bacteriologists and even all the biologists. The doctrine has become universal. It is one of the great aspects of the study of life and one of the most eminent titles of glory in French science. And even in this city of New York, one of the most fecund scientific institutes, the Rockefeller Institute for Medical Research, is, after all, only a large Pasteur Institute. The importance and fecundity of Pasteur's work evidently manifest themselves right here, then, in a patent and permanent fashion.

---

[1] Dr. Émile Roux died November, 1933, at the age of eighty.

March 22, 1933.

Sixth Lecture

CONTEMPORARY FRENCH SCIENCE

———

Mathematics.—Physics: Radio-activity: H. Becquerel,
M. and Mme. Curie.—Undulatory mechanics:
M. Louis de Broglie. — Chemistry. — Natural
sciences.

———

After having run through the nineteenth century as
a whole, it remains for us to glance at contemporary
French science.   I shall consider as such that which
corresponds to the period of activity of the men of
my generation, the one through which I, myself, have
lived.   It includes approximately the last forty years,
those which conclude the nineteenth century and what
has actually elapsed of the twentieth.   Here, even more
than before, appears the artificial character of these
divisions of time, and in the two preceding lectures it
has already been necessary to deal with men whose
career and whose work have been prolonged more or
less beyond 1900.   Scientific generations inevitably
overlap one another and their ideas do, also.   The only
logical subdivisions of the history of sciences would
be those which correspond to a renewal of ideas by
great discoveries, which modify the general orientation of research.   At any rate, this does not occur
simultaneously in all the sciences.   In the actual case,
the period considered corresponds fairly well with a
natural phase in physics and in biology.

More than for the preceding periods the difficulty
arises of making a choice among the works.   It is a
very delicate matter to speak of those who are still

living, or to pass over their names in silence. In the little time at my disposal, I must limit myself on one hand to major groupings; on the other, to some special works, the import of which has been particularly widespread; and that, all the more because, contrary to the seventeenth, the eighteenth and the beginning of the nineteenth centuries, scientific research is no longer restricted to a small number of individuals, but, thanks to the existence of manifold institutions and laboratories, the number of research workers is growing greater and greater.

Mathematics.

It is particularly difficult to speak of contemporary mathematics, for its character, in some way hermetical, is accentuated in proportion as it progresses. It is no longer accessible except to the small number of initiated people who devote themselves to it, and even those could reveal the content of it summarily to the public only with difficulty. On an occasion like this, then, one can only present the names of the men which the expressed opinion of their peers designates. Nobody will contest, in any case, that the French mathematical school of our day possesses a vitality and a brilliancy which are as remarkable as ever. The causes which had assured this result in the past continue to exist. The selection of mathematically talented students remains one of the most pronounced and happiest characteristics of the system of French secondary education. The great scientific schools, the *École Polytechnique* and the *École Normale,* in the first place, always exercise the same attraction over youth; the high level of the programs drawn for the entrance competitions, the large number of candidates make a remarkable selection possible. However, for almost half a century, the *École Normale* has been tending to become the principal place for scientific vocations. The *École Polytechnique* is continuing, without a doubt, to attract each year an elite which is as well endowed as ever; but, after school days, economic conditions, the brilliant prospects open until very recently to engineers in the business world, have certainly dried up, in a way, the current which formerly attracted some of the most original minds to pure science. One scarcely sees any more such examples as those of Gay-Lussac, Arago, Fresnel, Sadi Carnot, or Le Verrier, who, in their leisure time between duties, devoted themselves to scientific meditation. We must also add that scientific research to-day is hardly adaptable any longer to a division of activity; it requires, in general, the total devotion of one's personality, and the special resources of the laboratory.

At the end of the nineteenth century, Henri Poincaré was recognized as the first of the mathematicians of his time. He disappeared prematurely in 1912, in the midst of his productive activity. Some of his condisciples and emulators are still at work or have just barely passed away, such as P. Appell (1855-1930) and G. Koenigs (1858-1931). M. Émile Picard, who, from the very beginning, affirmed his mastery over very varied domains of mathematics, belongs to the same generation, as does also, M. E. Goursat. Separated by a more or less marked interval, a numerous series of eminent mathematicians follows them, among whom must be cited MM. P. Painlevé, J. Hadamard, E. Borel, E. Cartan, J. Drach, H. Lebesgue, E. Vessiot, P. Montel. Even this generation has already lost some of its representatives of great worth, such as P. Boutroux, R. Baire (1874-1932), Faton (1878-1929). To the preceding names should be joined those of younger men, as MM. Denjoy, Julia, etc. I cannot attempt to characterize the work of each and shall limit myself to merely referring to the new edition of *La Science Française*, which is going to appear.

In the field of rational mechanics, inseparable from mathematics and from physics as well, we would again find the names of many mathematicians mentioned above. An eminent and special place belongs to the work of J. Boussinesq (1842-1929), who is responsible for important advances in theoretical hydrodynamics, now finding their verification in the experimental researches of M. C. Camichel. Another physicist, M. M. Brillouin, has likewise occupied himself in a large measure, and with great success, with problems related to mechanics, such as those of elasticity. The mechanics of fluids, much in the forefront to-day because of the studies necessary for aviation, has recently given rise to considerable and remarkable research work. I am omitting the entire field of the applications of mechanics in which France has not lacked engineers, combining a very extensive theoretical knowledge with the keenest understanding of practical uses.[1] In this respect, I shall limit myself to recalling the name of Rateau (1863-1930), who disappeared recently and quite too soon, having been an initiator and a great master of everything pertaining to turbines, the invention of which, about a century ago, was already due to a French engineer, Fourneyron.

Physics.

Whatever the present development and the importance of the progress of mathematics in France may be, the contemporary period has no marked individuality in relation to those which have preceded it. It is the direct continuation of the nineteenth century.

The case is not the same for physics, which, pre-

---

[1] The art of engineering thus received valuable help from the graphic method in calculations, conceived and developed by M. M. d'Ocagne; as a whole, such methods constitute nomography, the application of which yields fruitful results in many diverse branches of experimental sciences.

cisely, has been passing, for about forty years, through a phase of almost total renewal. We still frequently receive to-day, from the lips of men who survived the generation that worked in the second half of the nineteenth century, the very significant piece of information that, around 1880, physics appeared to the majority of its devotees as a science which had been completed along its main lines. This state of mind was perhaps more marked in France, where admiration for the work of Fresnel, Ampère, Carnot, and their disciples continued. Nevertheless, a real revolution was in the making; its forerunner was the English physicist, J. C. Maxwell; his electromagnetic theory of light, a construction as yet purely theoretical and mathematical, tended to closely unite the two fields of optical and electrical phenomena, in the way Ampère had formerly allied magnetism and electricity. This theory took the form of concrete reality in 1888, thanks to the experiments of the German physicist Hertz, who revealed the propagation of electromagnetic waves corresponding to Maxwell's conception. Since then, that order of phenomena was to play a signal part in both theory and applications. Concerning the latter, among other consequences, one of the results has been that marvelous application, wireless telegraphy, and I shall recall that one of the first major steps along this path was Branly's discovery of the limature coherer in 1890. Among those who, after this, contributed most to the progress of this technique. we must not forget the name of General Ferrié (1868-1932), who has just recently passed away.

But, what was going to profoundly renew physics. was the concrete study of matter on the atomic scale, which, to-day, is the basis of research for all phenomena and for the conception of the whole of the universe. Contemporary physics can thus be called atomic physics, and it has even shorn the term of atom of its etymological meaning, since it has resolved the atom into a world of elements, at the limit of which matter is reduced to electric charges.[1]

French physicists and chemists, as a whole, had been attracted in only a middling degree, in the nineteenth century, by atomic conceptions. We have seen, for chemistry, that the atomic theory which proved so fruitful and which had been conceived in France, had met with tenacious opposition there from incontestably eminent minds. This was certainly due, in part, to a positivist turn of mind, which adhered to the doctrine of not going beyond the immediate facts known from experience, and which felt a repugnance against treating as a reality a whole system of representation, coherent, perhaps, but inaccessible to observation. The molecular and atomic conceptions, originated in chemistry, were introduced into physics and particularly into thermodynamics by the kinetic theory of gases, in which the classical laws relative to pressures exerted on the walls were conceived as the mass result of the movements and elementary collisions of the individual molecules. These molecules, inaccessible to direct observation, appeared as imaginary beings. We find a particularly well defined expression of distrust in respect to these hypothetical conceptions in the works of a physicist who was a great master of theoretical physics and particularly on thermodynamics, P. Duhem (1861-1916). This feature is revealed in all its distinctiveness in the following quotation[2] in which he declares that he refuses to treat "la physique théorique à la façon des cartésiens et des atomistes. Les corps que les sens et les instruments perçoivent, on les résout (dans cette conception) en corps immensément nom-

---

[1] For a synthetic exposition of these conceptions, see the book of J. Perrin: *Les Atomes* (Paris, *Alcan, Nouvelle bibliothèque scientifique*).

[2] Taken from *Histoire de la Physique,* by Ch. Fabry, p. 392.

breux et beaucoup plus petits, dont la raison seule a connaissance; les mouvements que l'on observe, on les regarde comme les effets résultant des mouvements imperceptibles de ces petits corps. . . . Ces corps, ces mouvements sont, à vrai dire, les seuls mouvements réels. Lorsqu'en les réunissant d'une manière convenable, on les a reconnus capables de produire des effets d'ensemble pareils aux phénomènes observés, on déclare que l'on a découvert l'explication de ces phénomènes."[1] Physics for Duhem, was, on the contrary, an energetics, the principles of which "ne se posent nullement en révélations sur la nature de la matière. Ils ne prétendent rien expliquer. Simplement, ils se donnent pour règles générales, dont les lois constatées par l'expérimentation sont des cas particuliers."[2] That is a profession of faith somewhat "superpositivistic," whatever the road traveled over in the course of more than half a century and the tremendous difference in the temperaments and philosophical opinions of Duhem and Auguste Comte. Made at a time when atomic physics took its flight, it still illustrates the deeply positive stamp of French science of the nineteenth century.

Now, all this world of imaginary beings, atoms and molecules, concerning which Duhem proclaimed his

---

[1] He refuses to treat "theoretical physics in the way the cartesians and the atomists do. The bodies, which the senses and instruments perceive are resolved (in this conception) into immensely numerous and much smaller bodies, apprehended by reason alone; the movements that one observes are regarded as the effects resulting from imperceptible movements of these little bodies. . . . These bodies, these movements are, really, the only real movements. When, by reuniting them in a suitable manner, one has recognized them capable of producing effects of a whole, similar to observed phenomena, one declares that the explanation of these phenomena has been found."

[2] The principles of which "are by no means presented as revelations of the veritable nature of matter. They do not claim to explain anything. They simply give themselves as general rules, of which laws verified by the experimenter are particular cases."

skepticism, has become, in the last decades, a real and tangible world, whose reality our senses register indirectly and partially at least. We perceive molecules and atoms now, and in the atom we discern, with its central nucleus and its train of electrons, an entire world. Atomism of which the Greek philosophers had had a general idea, atomism which chemistry, and particularly organic chemistry, had been brought to consider as an essential postulate and which had been introduced into physics by the kinetic theory of gases, has become a positive reality in which the entire universe is recapitulated. One cannot insist too much upon the importance of these new data, upon the ample change they have brought about in scientific thought. For physics, they have the value of a revolution, which, without destroying previous acquisitions, brings about decisive illuminations, extends the boundaries of knowledge, and unifies it. This renovation in physics is a work accomplished by the joint efforts of different countries, a work in which numerous physicists have taken part. We have to examine the rôle played in it by French physicists.

One of the first significant contributions to this structure was a work by M. J. Perrin, who entered scientific life precisely at the moment when the new ideas were about to assert themselves. He made at his outset in 1895, an important discovery. By means of a simple and elegant experiment, he proved that the cathodic rays, produced in gas rarefied by electrical discharges, behave like real material projectiles, even though they are formed simply by negative electricity launched at a very great speed.

Another phenomenon also presented itself as a materialization of molecular movements: it was the Brownian movement, known by all microscope observers. When a liquid holds little particles in suspension, one can always see in the microscope these little particles in a perpetual state of agitation devoid of any definite direction or regularity. The eminent physicist Gouy (1854-1926), in 1888, was led to attribute this motion, with its features of complete irregularity to the incessant bombardment that the particles in suspension must receive at every movement from the molecules constituting the liquid, and to find thus in the Brownian movement a proof in favor of that of the molecules in the kinetic theory of gases. This interpretation is now classic. Some years later, J. Perrin, taking hold once more of these ideas and applying them to colloidal solutions in which the colloid particles offer an enlarged image of molecular phenomena, was able to draw from them experimentally a numerical count of the molecules in a given volume (what is called Avogadro's number) the result of which agrees very satisfactorily with the results deduced from theoretical considerations.

But let us return to 1895, to the time of Perrin's experiments on cathodic rays. Almost simultaneously, an extremely important discovery in a neighboring domain was made in Germany by Röntgen, that of the X rays. These rays were destined to be one of the most important instruments of research in modern physics: their nature was at first mysterious, as the name given to them indicates. It did not take long to realize that they were radiations of the same nature as those of light, but with much shorter wave lengths. It is their diffraction in crystals which permitted the recognition of their nature, thanks to the discoveries of the German physicist Laue. I must note in passing that the work of Laue has produced a decisive material verification of the theories formulated on the structure of crystals in the middle of the nineteenth century by the French crystallographer Bravais. Experimentation on X rays was immediately undertaken in every country. In France, it was particularly pushed forward in the special laboratory devoted to it by M. Maurice de Broglie, whose research work, as well as that of his collaborators, namely M. Thibaud, Trillat, and Dauvillier, has produced in this domain numerous and important results, especially in the spectrography of rays, their photo-chemical effects and the production and study of rays whose wave lengths and peculiar qualities constitute a transition toward ultra-violet rays and the visible spectrum. Thus the X rays are now definitely connected with luminous radiations.

Radio-activity: H. Becquerel, M. and Mme. Curie.

Of all the discoveries of that period, that which was the most unexpected and of greatest consequence was radio-activity.   It has enabled us to witness phenomena of an absolutely new order, the disintegration of atoms, and has shown us, in some cases, very limited in number, it is true, a reality corresponding to the old dream of the alchemists, the transmutation of chemical elements.   We know to-day, in fact, that in a radio-active substance, following a definite rhythm, and outside of the action of all external physical forces known at present, atoms disintegrate; they pass through a series of temporary and regular phases, which represent as many unstable chemical elements, in order to arrive finally at a stable state, which is the atom of another simple substance.   The process consists of successive transmutations of the primitive atom, degradations by virtue of the loss of certain ones of its constituents.   Some of these transformations are very rapid and are expressed by perceptible effects after months, days, hours, minutes, and seconds, as the case may be; others, on the contrary, require centuries, or even thousands of centuries.   It is, then, a case of phenomena taking place on the atomic scale, phenomena which restrict within bounds, limited and precise, moreover, the principle of the conservation of matter, such as it had been conceived since Lavoisier; these transformations take place with an extremely great release of energy which manifests itself in the form of an emission of projectiles of different categories, shooting forth at a tremendous rate of speed and made up of the particles lost by the atom. There we have, then, a class of phenomena of which we formerly had no idea, and which changes our entire conception of matter and the universe. As far as terrestrial conditions are concerned, there is only a small number of elements presenting these degradations. But it is not impossible to believe that we have here a general characteristic, capable of manifesting itself under conditions other than those of the surface of our planet, and that the atoms of the different chemical elements are more or less complex bodies made up of the same elemental particles joined together in varying numbers and in accordance with different types of structure. Thus, we come to a modern form of the conception of the unity of matter, a unity which the philosophers of ancient Greece had already surmised, and which, in the nineteenth century, presented itself to the minds of certain chemists. At the beginning of the nineteenth century, an English physician, who had never done any original work in chemistry, Prout, had the singular good fortune of formulating intuitively, in 1815, the idea that hydrogen is the universal constituent of matter. It is exactly at this conception that we are arriving to-day.[1] These few remarks show the tremendous import the discovery of radio-activity has had, and still has. Now, this discovery is French, and the circumstances under which

---

[1] The atom of hydrogen with one positive charge constitutes a proton. The nucleus of an atom is made up of protons and electrons. For each chemical element, the number of electrons rotating around the nucleus is fixed. (Atomic number.)

it was realized are very significant as regards the history of science.

Its starting point lay in a suggestion that H. Poincaré, who took as much part in the progress of ideas in physics as in mathematics, made at the beginning of 1896, at the time when Röntgen discovered the X rays, and when neither the nature of these rays, nor their production was known. Poincaré put forth the hypothesis that the latter could be linked up with the fluorescence of glass struck by cathodic rays, a hypothesis which was not to be subsequently verified. But this suggestion led H. Becquerel,[1] who occupied himself at the time with phosphorescence, to look for a contingent production of X rays on uranium salts made phosphorescent by the action of light. At first, it seemed as though experiments verified the prevision. The isolated uranium salts made an impression on a photographic plate protected by black paper. But Becquerel was not long in finding that the phenomenon was produced just as well, even when the salts had not been exposed to light, and that it was absolutely constant and common to all uranium salts, depending only on the presence of uranium, whatever the state of its combination. He concluded from that that there was in them a radiation of a new nature, belonging exclusively to the uranium atom. That was, in fact, the beginning of radio-activity. This radiation received at that time the name of Becquerel's rays. Becquerel continued his study of it and found in it important properties, notably the distinction of rays of different categories. Becquerel's investigations,

---

[1] Henri Becquerel (1852-1908), son of Edmond Becquerel (1821-1891), and grandson of Antoine Becquerel (1788-1878), both physicists of high standing, whose names have already been mentioned. The work of Antoine Becquerel bears especially upon electricity and magnetism, on piles and electrolysis in particular. Edmond Becquerel has left important works, namely on ultra-violet rays and phosphorescence.

completed by those of M. Villard, ended in the distinction, thereafter classic, of three categories of rays, α, β and γ, a distinction parallel to the one furnished by the production of X rays (cathode rays, anode rays [Kanalstrahlen], X rays).

A more exact study of the radiation of uranium and its sources was essential; at the same time, it was necessary to see whether phenomena of the same order could not be produced with other chemical elements. This Mme. Sklodowska-Curie undertook in Becquerel's laboratory, and the results shortly obtained by her led her husband to participate in this research.

Pierre Curie (1859-1906) had already revealed himself as an eminent physicist by means of his various accomplishments. He had begun with remarkable studies of the infra-red spectrum, continuing the work of the young physicist Mouton, who had died almost immediately after writing thesis for the doctorate. With his brother Jacques Curie, Curie had next made the important discovery, and also a study, of the piezo-electricity of quartz;[1] finally, he had done some very fine work on magnetism and was to add other very profound studies on symmetry in physical phenomena. Unfortunately, Pierre Curie's career was destined to be prematurely broken off. In 1906, before his forty-seventh year was ended, he passed away, run over by a truck in the streets of Paris. The works of M. and Mme. Curie are remarkable not only because of the final importance of the results obtained, but also because of their skill in analyzing entirely new and complicated phenomena, which they knew how to interpret and clarify remarkably well. Naturally, it is possible to indicate here only the most essential points.

Mme. Curie began by looking among all the chemical

---

[1] Production of the two electricities of contrary signs on a quartz crystal suitably cut and subjected to torsion forces.

elements for those which produced a radiation similar to the one produced by uranium; that was the case with thorium alone. She finished proving that in the case of uranium, this radiation is a property of the atom, independent of compounds and external circumstances. One of the properties of the radiation was to produce electric conductivity in gases and to ionize them, an effect which furnished a means of measuring exactly, by the degree of ionization, and particularly with the aid of Curie's piezo-electric quartz, the intensity of the radiation. It was thus recognized that the radiation of uranium is constant. When she studied the radiation of

Pierre Curie

uranium ores, Mme. Curie was not long in finding out some where it was definitely more intense than the radiation of uranium itself. M. and Mme. Curie then supposed that it was due to another substance still unknown. By proper treatment of the ore, they were able to concentrate the treated parts in which this special radio-activity manifested itself, and obtain radiations which became more and more active. This work of purification subsequently proved to be a gigantic one, and it was truly a heroic deed to attempt and pursue it with tenacity, for it was a question of operations on an industrial scale made in a tiny laboratory, completely devoid of the appropriate resources and apparatus. In the course of this work, Mme.

Curie was soon led to distinguish a radio-active substance whose chemical properties resembled those of bismuth, and whose radiation was much more intense than that of uranium. She called it polonium.[1] But at the same time another substance revealed itself in the same mineral (the mineral studied was the pitchblende of Joachimstahl, in Bohemia), even more radio-active than polonium, chemically related to barium, and carried away with it into the purification processes. It is that substance that in 1898, Mme. Curie and M. Bémont described as radium. Above a certain degree of concentration, polonium and radium were distinguished and characterized by their particular spectral stripes, thanks to Demarçay. Radium, progressively purified to a bromide state, has shown a radio-activity 1,800,000 times as great as that of uranium; in 1910, Mme. Curie obtained it in its metallic state. Now, then, it is an isolated chemical element whose atomic weight[2] is extremely great (225), not much less than that of uranium (240), or than that of thorium (232). Later, M. Debierne identified in pitchblende another radio-active element, actinium, derived, like radium, from uranium.

But these substances, and especially radium, gradually revealed disconcerting and enigmatical properties, which were in disagreement with all the phenomena previously known. It was soon found, in fact, that all the objects in the laboratory possessed radio-activity, temporarily at least, and with varying intensities. M. and Mme. Curie called the phenomenon induced radio-activity, and studied it. Finally it was recognized by Rutherford that this is due to a gaseous substance, to which the name radium-emanation (radon) was given.

---

[1] One was subsequently led to believe that polonium might be a transitory product resulting from the disintegration of radium.

[2] The radio-active elements are among those whose atomic weights are the greatest.

Radon, an autonomous chemical element destroying itself rapidly, is a product of the disintegration of radium; this substance finally produces helium, as Sir William Ramsay and F. Soddy have proved in England. Another riddle put by radium was a continuous and considerable release of energy. After thinking that this energy could proceed from the collisions of an external radiation, Curie concluded that it was liberated by the disintegration of radium. Those were the principal phenomena, entirely without precedent, discovered through the study of radium.

From 1896, the date of Becquerel's discovery of the radiation of uranium, to 1906, the time of Curie's death, M. and Mme. Curie and their collaborators were far from being the only ones working on these problems. Because of their tremendous interest and their mysterious character, these problems attracted groups of research workers in every country, and it is the English physicists and chemists Sir J. J. Thomson, Rutherford, Sir William Ramsay, Soddy, etc., who, together with the French group, M. and Mme. Curie and Debierne, brought forth the essential points of the results. What was learned in the following years was due, even to a greater extent, to the work of a great number of scientists, but it can be said that all the surprising novelty of this chapter in science goes back to H. Becquerel and M. and Mme. Curie.[1]

It is not necessary to insist on the extent of the consequences of radio-activity as regards physics. It has permitted us, for the first time, to study the true atomic properties and to witness the destruction of atoms. The different rays themselves have become, in the hands of physicists, a powerful agent, projectiles possessing a formidable amount of energy. Therefrom

---

[1] Becquerel and Curie were also the first to suffer from cases of radiodermite, due to their handling of radio-active products at a time when the effects on the organism were still unknown.

proceeds almost all that is essential in contemporary physics, the development of which we see continually broadened by this new order of phenomena. Among the young French physicists now devoted to these studies, let us cite, besides M. Debierne, Mr. Holweck, M. and Mme. Joliot-Curie, M. E. Perrin, M. P. Auger, etc.

Present-day physics has received other contributions from France. In the field of magnetism, wherein P. Curie had accomplished some fine research, M. Pierre Weiss, now professor at the Faculty of Strasbourg, has to his credit considerable work, from the theoretical as well as experimental point of view. This work has formed the basis for a new branch of chemistry—magneto-chemistry—or the study of the relationships between chemical composition and magnetic properties.

The application of interferences and wave-length measurements to the study of optics and to highly precise mensuration in which Fizeau had distinguished himself, has brought about a considerable body of research by J. Macé de Lépinay, Pérot, Ch. Fabry, H. Buisson. The results have contributed toward the establishment of an international system of wave lengths, accepted as the universal basis for standards of length, standards which are nondestructible, free of variation and accident, and to which metric units are linked. Applications of spectroscopy to chemical analysis have been carried to a high point of perfection by Eug. Demarçay, and by Arnaud de Gramont (1861-1923). We must also not fail to mention M. A. Cotton's important researches on circular dichroism and on magnetic birefringence. These have led to the construction of an electrical magnet, at present the most powerful in the world, with which numerous problems can be studied, including the problem of changes in the properties of matter in an intensely magnetized field. M. Langevin is one of the physicists who to-day command the theoretical conceptions of atomic physics, and to whom we owe various and very remarkable experimental achievements like the use of piezo-electricity in depth sounding at sea by means of supersonic waves.

The name of M. André Blondel will remain connected with the adaptation of the modern theories of physics to many applications: namely, in electrodynamics (oscillographs, radiotelegraphy, radiogonometry, radiophares), in optics (photometric unities, range of short signals and of searchlights, progress in lighting and lighthouses), in mechanics, finally in the use of the interferences for the acoustic and hertzian signals.

M. Louis de Broglie's Theory of Undulatory Mechanics.

In the field of phenomena considered in regard to the atomic scale, a new science has emerged, since 1924, from the work of M. Louis de Broglie, the science of undulatory mechanics, which corrects, according to this scale, the Newtonian mechanics, now looked upon as an approximation. These studies have also removed a disturbing contradiction in relation to the fundamental conceptions of optics. A very large place in contemporary science has been taken up by the theories of Planck, formulated about 1900, which make energy a noncontinuous magnitude varying according to units, or quanta. This theory has been extended to light, which is likewise conceived as formed of light-quanta, or photons; we have there a corpuscular conception which takes us back to the emission theory, abandoned, since Fresnel, because of its inability to explain the phenomena of interferences and diffraction. We were faced, therefore, with a fundamental contradiction. It has been removed by M. L. de Broglie, who has attached to all corpuscular movement a wave of appropriate frequency. The consequence of this conception is that we must obtain for movements of electrons phenomena of diffraction, which result from the accompanying waves and interferences. M. L. de Broglie himself pointed out that this conception should be experimentally verified by letting fall bundles of electrons of known velocity on crystals, where the waves would be brought to view by the phenomena of crystalline diffraction discovered by Laue in 1912. The experimental verification he thus foresaw was decisively accomplished in New York, in 1927, by Davisson and Germer. It has corroborated not only the reality of the undulatory phenomenon, but also the value of the wave length calculated *a priori* by M. L. de Broglie. Other verifications have been performed independently since then. This conception of luminous phenomena carries into mechanics, and leads in regard to mechanical actions on the atomic scale, to a new theory of mechanics, undulatory mechanics, which bears the same relationship to classical mechanics as undulatory optics does to geometrical optics.[1]

---

[1] M. L. de Broglie has been effectively guided by the analogy between the equations of analytical mechanics and those of geometrical optics. Maupertuis's principle of least action holds in the former the place that Fermat's principle of minimum time of propagation holds in the latter. The possible trajectories of a corpuscle identify themselves with the light-beams in the optical field.

If h is Planck's constant (quantum of action, $h = 6.55 \times 10^{-27}$), *nu*, the frequency of radiation, the energy of a corpuscle $E = h\nu$, (the X rays behave like corpuscles endowed with an energy E). Let v be the velocity of a corpuscle: the quantity of motion being $g = mv$, the wave length associated with the movement of the corpuscle will be: $\lambda = \frac{h}{g} = \frac{h}{mv}$.

We see, or at least we glimpse from these summary considerations, the novelty and breadth of M. L. de Broglie's conceptions and we understand that henceforth they have an experimental basis. For this reason, they hold a leading place in the science of the last eight years. They have renovated most of the fundamental problems of mechanics and optics considered from the point of view of the atomic scale. It can be said without exaggeration, that they form one of the essential aspects

Louis de Broglie

of the contemporary scientific mind in one of its loftiest provinces, since they deal with the elementary developments that go into the making of the universe.

Chemistry.

As a result of the preceding facts, physics has, at the present time, undergone a deep and fruitful transformation. It is now, as its name implies, the science of Nature. It reduces the whole universe to the atom and makes atomic phenomena the universal basis for science.[1] For this reason, discoveries have followed rapidly in the field, new conceptions replace old ones, and sometimes clash in a rather alarming way. In comparison with physics, the other sciences appear to be somewhat in a state of stagnation. In chemistry,

---

[1] See the work already mentioned: J. Perrin, *Les Atomes.*

for instance, the only facts which have a quality of newness are those connected with radio-active bodies, facts which we might say belong rather to the field of physics; we have already encountered them. Nevertheless, the transmutation of radio-active substances, as well as the correlative variations of the mass are breaks in the absolute dogma of the "conservation of matter" such as one would not have dared to imagine. But their bearing is strictly limited and the laws of chemistry survive without reservations, outside of the very special case of radio-active substances.

We cannot think of entering into an investigation of the innumerable specific results which French chemists have obtained in the course of the last decades, though many be of importance to specialists. I shall limit myself to mention very briefly the work of M. P. Sabatier and his collaborators, Sendérens and Mailhe on catalysis promoted by metals in a divided state (platinum, iron, cobalt, nickel, copper) ; this work has furnished highly fruitful methods to organic chemistry. Outside of the actual achievements which they have been responsible for, these methods are of considerable interest in that they secure numerous reactions to relatively low temperatures and, it is precisely through catalytic actions at normal temperatures that the greater part of chemical transformations occur in organisms. During the years preceding his death, Moureu (1863-1929), had also discovered catalytic actions in oxidations or in conditions which prevent oxidations (antioxygen substances).[1]   M. Grignard has contributed a powerful means of synthesis to organic chemistry through his discovery of the organic compounds of

---

[1] These substances are first of all phenols. Very small quantities of hydroquinine, for example, prevent the oxidation of benzol aldehyde in contact with oxygen, or, at least, slow it enormously. From this fact, applications may be derived in regard to the processes of oxidations in the organism.

magnesium. In the field of biological chemistry, M. Gabriel Bertrand has made remarkable discoveries on widely different subjects, such as the discovery of oxidizing diastases (like laccase), and the mechanics of their action. They constituted the first example, followed later by many others, of a soluble ferment which is not made up of a single agent, but of a pair of substances (ferment and co-ferment); in the present case, the co-ferment which functions as catalyst is a metal in minute but constant quantity, manganese. In other biological actions like those of the venom of serpents, which C. Delezenne has admirably studied, the co-ferment is made by zinc. M. G. Bertrand's discovery certainly has a wide import. Moreover, with his finished skill as an analyst, M. Bertrand has established himself as a leader in the study of chemical infinitesimals. In particular, he has revealed the presence in organisms of a large number of chemical elements which one did not suspect were there. Among the eminent French chemists who passed away more or less recently, I shall recall the names of Haller (1849-1925), Bouveault (1864-1909), L. J. Simon (1867-1925), in organic chemistry; that of Maquenne (1853-1925), in biological chemistry and in vegetable physiology (the osmotic pressure of cellular juice, the respiration and chlorophyllous activity, the physiology of grains, the part of chemical infinitesimals in vegetation), that of Bourquelot (1851-1921), who, with his collaborators, enriched in an important way the chemistry of glucosides and of their diastases. Among the living we could not ignore, besides those we have already mentioned, the names of MM. Béhal, Urbain, Delépine, Matignon, etc.

Chemistry cannot be detached from its applications, and France has not ceased to provide at least the starting point and often the full realization of important new industries. If the chemistry of coloring matters has developed fully in Germany in the second half of the nineteenth century, we could not forget that it was first achieved by the French, through the work of Verguin, Coupier, Lauth, Roussin, etc. . . . who dis- covered some of the basic dyes (fuchsine, azo-dyes, dyes derived from naphtalin). At the present time, the liquid-air industry has received great impetus and enrichment from M. Georges Claude, both in methods and in products, whether it be oxygen or rare gases like neon, crypton, etc. One cannot utter M. Claude's name without recalling his daring experiments in the use of cold water from under-sea depths in tropical regions, as a cold source of energy for a machine to produce steam and power. In keeping with Carnot's principle, these experiments, whose development is at present held in check by economic difficulties, are very likely the prelude to a vast utilization of an unlimited and, up to the present, functionless force of nature, a utilization which would considerably change the con- ditions of human life on the borders of the tropics. The artificial silk industry also had its origin in France with Hilaire de Chardonnet (1839-1924).

Natural Sciences.

In considering the recent development of natural sciences, one is faced with the same difficulty as in the case of chemistry. There have been no startling discoveries to change its aspect, except the study of the problem of heredity which forms one of its new and fully developing branches, genetics; but genetics has really been little studied in France, although some of its most distinguished precursors came from there.

Concerning their descriptive aspect—systematic in both realms, comparative anatomy, embryology, histol- ogy, cytology—the particular studies, whatever their value may be, are only stones which go to make up the building as a whole, which arises gradually and whose outlines are henceforth fixed. We could not, then, go into details in this respect.

It is a necessity for the scientific mind not to be content with mere description, but to group facts and arrange them in a hierarchical order, connect them with principles, and thus obtain an explanation of reality by giving, with more or less precision, an account of the manner in which it was produced. Thus has evolution remained, since the middle of the nineteenth century, the central problem and the far pole of biology, just as the mechanism of the development of the individual, beginning from the egg, is its near pole. Biologists attempt to penetrate the determinism of both processes. But, in view of the complexity of the phenomena, and, in so far as evolution is concerned, in view of the fact that it does not belong to the present, the mind is constantly obliged to go beyond the strict limits of facts based on observation or experiment and to combine them through speculative conceptions of a more or less general order, through working hypotheses which are no longer in the realm of positive science. It will not be surprising, after all that has been previously said, that French biology maintains, in this respect, particular discretion, which proceeds from the need of the French mind for precision and clearness. When we examine the account of the last three quarters of a century that have passed since the impetus given by Darwinism, the sum total of French research work may seem a bit slender with reference to what other countries have to offer, especially Germany, where there flourishes a taste for speculative hypotheses, used as framework and support for specific facts. Set according to the hypotheses thus devised, studies multiply, conjectures increase and become confused, words are created, which often seem to be a reality. But when, drawing back to a certain distance, we consider the results, if we cannot deny that this tendency is a generator of work, we must declare that the substantial, ultimate results are not always in keeping with the effort made. At the end of forty or fifty years, many immense systems, based on purely speculative conceptions hastily formed or too boldly generalized, remain no more than burying grounds for ideas. The men of my generation witness an almost complete recasting of the general theories and conceptions that were in favor at the time of their first attempts and the aging of those that seemed to have the most promising future. Undoubtedly, progress cannot go on without loss, but is it necessary for it to discharge its obligations by such a waste of work, which, seen in retrospect, appears to a great extent ineffectual because inspired by directive ideas with an insufficient basis?

The French mind, because of caution which may sometimes be extreme, does not readily commit itself to speculative conceptions and prefers to abide by the immediate data provided by observation and experiment. If France's output in biology and, more particularly, in zoölogy, has been limited, it has, nevertheless, in recent times, contributed to science a good number of firmly established results. It will suffice to recall a few names, among many others, as those of Alfred Giard, Y. Delage, Math. Duval, E. Maupas, L. F. Henneguy, Albert Pézard, who all died more or less recently, and, among the living, E. L. Bouvier, P. Marchal, F. Mesnil, M. Caullery, L. Cuénot, L. Léger, O. Dubosq, Ch. Pérez, E. Guyénot, Ed. Chatton, P. de Beauchamp, etc. . . . In more than one case, French observations or experiments have been at the basis of investigations carried on over the whole extent of the scientific world. To cite but two examples, the discovery of the hæmatozoon of malaria, a discovery so rich in results, was the work of a French observer,

A. Laveran (1845-1922), in 1880. Experimental embryology numbers among its first creators, L. Chabry (1855-1893), who clearly formulated almost all its problems and worked on them with original and extremely elaborate experimental methods and instrumentation. In the realm of microscopic anatomy and histology, Ranvier's creative tradition has been brilliantly continued through such men as A. Prenant (1861-1927), Laguesse (1861-1927), Déjerine (1849-1917), P. Bouin, A. Ancel, Cl. Regaud, etc.

In botany, too, French science has recently brought important contributions to the various branches of this science: to cytology, through L. Guignard (1852-1928), Al. Guilliermond and his pupils; to cryptogamy, through J. Costantin, L. Matruchot (1863-1921), P. A. Dangeard, C. Sauvageau, to anatomy, through G. Chauveaud, to the study of the various problems of the relations between the plant and its environment, through G. Bonnier (1853-1922), Noël Bernard (1874-1911), L. Mangin, to the study of flora, mainly in the French colonial domain, through A. Lecomte, Aug. Chevalier, R. Maire, H. Jumelle, H. Humbert, Perrier de la Bathie, etc.; to genetics, through L. Blaringhem;[1] to physiology, through M. Molliard, Devaux, etc. Animal and general physiology has continued during the last decades, to produce eminent representatives in France. M. Ch. Richet has opened two large new channels by formulating, at the same time, the idea and the first principles of serotherapy and anaphylaxy. M. A. d'Arsonval has contributed brilliant, original creations to the realm of electrophysiology; C. Delezenne (1868-1932), who died very

---

[1] In the practical application of genetics, are to be mentioned the very important work of M. Schribaux on wheat and the making by M. Vialal, of hybrids between American and European vines which are able to be multiplied by slips. By those hybrids was gained the possibility of saving the French vineyard from the attack of phylloxera.

recently, leaves a series of studies of a profound originality, while stamped with the care of precision and experimental perfection, in the fields of the coagulation of the blood, of the conditions of activation of the pancreatic juice, and of the properties and composition of venoms. E. Gley (1857-1930), made extended studies of internal secretions, L. Lapicque, of the physiology of the nerve, where he introduced and generalized the standard idea of chronaxie. It is, moreover, to the purpose to recall with a word, the studies of J. Lefèvre on biological energetics, of Terroine, of A. Mayer; of Wurmser on metabolism, of M. Nicloux on physiological chemistry; of Doyon, of Pachon, of Hédon, of Tournade on the various problems of the physiology of organs; those of Raphael Dubois (1849-1929), on animal luminosity, whose fundamental diastasic mechanism (luciferin and luciferase) he has made prominent, and on hibernation; those of Portier and Cardot on comparative physiology, etc.

A last branch of the group of natural sciences must not be omitted, for it continues to show great vitality in France: it is geology. We have already had occasion to recall the names of eminent, recently deceased French geologists and those of men whose career is now advanced: behind them a new generation continues the tradition, extending the scope of its researches more and more in the colonial domain of France.

Confronted with the impossibility of acquainting the reader, in a few words, with the various aspects of this many-sided development, I shall confine myself to referring him to the very well informed and methodical outline from the pen of M. Emm. de Margerie in the chapter devoted to geology in *La Science Française*. I shall simply point out the eminent position which M. A. Lacroix holds in the field of that science, on account of his masterful researches on volcanism, and the exceptional breadth and accuracy of his work on the structure of eruptive and crystalline rocks; in these studies, he has shown himself the worthy continuer of Fouqué and Michel-Lévy. His splendid monograph on the eruption of Mount Pelée in 1902, among other studies on volcanoes, and his vast *Minéralogie de la France et de ses colonies,* constitute two capital achievements of contemporary science. M. Lacroix is, truly speaking, even more of a mineralogist than a geologist. In the special field of physical mineralogy, I must mention among others, the names of MM. F. Wallerant, G. Friedel, Ch. Mauguin, who have made important contributions to their particular science, in the principal branches now being brought to the fore (liquid crystals, analysis of the crystalline structure with the help of X rays, etc. . . .).

Each science has its own history. The history of the sciences in itself tends, in our day, to become an individual discipline. France has made its contribution thereto, and in this respect I shall recall the names and works of Paul Tannery and of P. Dunem.

CONCLUSION

Having reached the end of this too rapid review of French science from the seventeenth century to the present time, it remains for us to set forth, by way of conclusion, its essential characteristics.

First of all, we can state that France has played one of the most important parts in the building up of modern science as a whole. This is in no way a depreciation of the work of other nations, for example, of England, where the scientific life of the last three centuries has been one of uninterrupted and prolific activity and where men of genius have at no time been lacking; it suffices to think of Harvey, of Newton, who is perhaps the prince of all modern science, of Faraday, Maxwell, and Ch. Darwin, to speak only of the dead. We do not underrate the work of Italy, which has been the herald of modern science, nor of Germany, where research has soared to such heights during the nineteenth century, nor that of Holland, where, in a narrow territory, so many first-rate men of science have constantly risen into prominence. Nor do we disparage the work of the Scandinavian countries, or that of the United States, where the scientific movement, originated in the middle of the eighteenth century with such men as Benjamin Franklin, has acquired tremendous breadth, especially during the latter half of the century; a country where science has at its disposal for research work material equipment in proportion to the economic resources of the land, and where men and discoveries correspond more and more to the scale of such an organization. Science becomes more and more every day the collective, and, so to speak, indivisible task of all the nations, since, at the present time, there is no notable discovery which does not provoke everywhere, the moment it is announced, investigations rivaling in enthusiasm and speed.

France can claim as her own a good part of the discoveries and great pioneers in this universal work.

To recall only the most illustrious names—and one is at a loss to limit the list—have we not met in mathematics Descartes, Fermat, Pascal, Laplace, Lagrange, Cauchy, Galois, Hermite, Poincaré; in physics, Pascal, Fresnel, Ampère, Carnot, and Curie; in chemistry, Lavoisier, Berthelot; in the biological sciences, Réaumur, Lamarck, Cuvier, Claude Bernard, and Pasteur, that is to say, in each science, great creators, most of whom have been founders of entire branches of science?

In the field of applied science, which, in the course of the period we have been studying, and especially for somewhat over a century has so profoundly changed man's material life and so prodigiously increased his power, the part of French inventions is no smaller. Without attempting a complete enumeration, it is sufficient to remember that several of the essential organs of modern technique were originated, complete in every respect, in France: the dynamo (Gramme), the electrical accumulator (Planté), the internal combustion engine (Lenoir), the turbine (Fourneyron), the automobile, which was created and perfected until it reached its final form; there, were conceived the use of water power (Bergès), the transmission of power at a distance through high-tension currents (Marcel Déprez), the refrigeration industry (Tellier), photography (Daguerre), cinematography (A. and L. Lumière), aërial navigation (Montgolfier), the dirigible (Ch. Renard), and under-sea navigation, which reached its definite stage with Laubeuf's submarine; all this, without taking account of a mass of instruments such as, in navigation, the gyroscope (Foucault), depth-sounding apparatus based on supersonic waves (Langevin). Besides, how many new ideas came to life in France and found elsewhere conditions favorable to their development! Papin's engine was the rough draft of the steam engine; the Marquis of Jouffroy d'Abbans was the father of steam navigation; Bourseul outlined the principle of the telephone, and Branly's coherer was one of the first stages in the creation of wireless telegraphy. There were made many other capital acquisitions, notably in the field of chemistry, not to overlook, lastly, all the applications of Pasteur's discoveries in the field of biology.

It can even be said that French science is distinguished for the number and value of its great discoveries rather than for its general mass of scientific production. The development of science can be said to result from two factors which in a way complement one another; on one hand, the great innovative ideas which open new paths; on the other, the patient and methodical work which uses them and prepares the discoveries to follow. Yves Delage liked to compare the first to rockets which suddenly illumine the dark sky and he would almost have liked to limit scientific progress to these sudden, unexpected, and sporadic events. In this way, modern chemistry emerged suddenly from Lavoisier's brain, electrodynamics from that of Ampère, thermodynamics from Carnot's and microbiology from the mind of Pasteur. French science is of worth more for its great discoveries than for the collective labor that extends them.

This fact hinges on a deeply rooted quality of the French mind, which is apparent in all phases of national life, the tendency toward individualism. The Frenchman does not possess the gregarious instinct; the individual is jealous, perhaps even too much so, of his independence and personality. He does not willingly let himself be absorbed into the team. There have been schools around the great masters of science, but their number has certainly been more limited than in other countries. Many masters have not especially sought to surround themselves with disciples. We can say that in France we have not fully benefited from average talents, because they are insufficiently organized under leaders and insufficiently disciplined. French science is above all individualistic.

If this inclination tends to preserve originality, it frequently has the drawback of scattering effort instead of concentrating it, and of breaking up resources to too great an extent. This criticism can certainly be made of the present condition of the French scientific world where there would be every advantage in lessening somewhat the number of laboratories and at the same time equipping them more efficiently. The slenderness of its material resources has been up to a very recent date and still is, in a relative way, one of the primordial troubles of science in France. Scientists of high merit and even men of genius have, almost without exception, accomplished their work with rather wretched equipment. It would be difficult to cite an important discovery which was made in a large laboratory. We saw Fresnel constructing with the help of a village locksmith the instruments which served in his crucial experiments on light. Ampère did not have much fuller resources at his command. Pasteur worked for a long time in an attic, Claude Bernard in a cellar, where his health was impaired. Until the end of the last century, properly equipped and endowed laboratories were rare indeed. It would be easy to assemble the cries of distress uttered by the most noteworthy scientists. "J'ai connu," says, for example, Claude Bernard,[1] "la douleur du savant, qui, faute de moyens matériels ne peut entreprendre ou réaliser les expé-

---

[1] "I have known," says, for example, Claude Bernard, "the sorrow of the scientist who, for lack of means, cannot undertake or achieve the experiments he imagines and is forced to give up certain investigations, or to submit his discovery in a sketchy state."
Rapport sur les progrès de la Physiologie en France (Exposition de 1867, p. 148).

riences qu'il conçoit et est obligé de renoncer à certaines recherches ou de livrer sa découverte à l'état d'ébauche." The majority of great discoveries have thus been brought into being in conditions of poverty, if not in misery, and if we must admire their authors all the more for this, there is no reason to pride ourselves on it. We can even, with Lippmann, reproach French society as a whole for not having taken enough interest in science, nor having been sufficiently aware of its importance. If this had not been the case, by the way, the material conditions of scientific research would have been improved earlier.

Repeatedly, in the course of the preceding account, I have stressed an invariable characteristic of French science, its essentially positivistic tendency. It is especially marked during the nineteenth century, but it already asserted itself in the seventeenth. Even a mystic mind like Pascal's proclaims the supreme authority of the experiment in the domain of science. Réaumur's is the typical positivist mind. Lavoisier in his *Traité de Chimie* writes, "Je me suis imposé la loi de ne procéder jamais que du connu à l'inconnu, de ne déduire aucune conséquence qui ne dérive directement des expériences et des observations. . . . Tout ce qu'on peut dire sur le nombre des éléments et leur nature se borne, suivant moi, à des discussions métaphysiques: *ce sont des problèmes indéterminés qu'on se propose de résoudre, qui sont susceptibles d'une infinité de solutions, dont il est probable qu'aucune particulière n'est d'accord avec la Nature.*"[1] One could not express better this positivistic conception of scientific research. Foreigners,

---

[1] "I have made it a rule to proceed always from the known to the unknown, to deduce no consequence which does not derive from experiments and observations. All one can say on the number and nature of the elements confines itself, in my opinion, to metaphysical discussions: *these are indeterminate problems that we attempt to solve, problems which are open to an infinite number of solutions, none of which, in all probability, are in accord with Nature.*"

at any rate, make no mistake about it.   The great
German chemist Liebig, contemporary and emulator of
J. B. Dumas, writes in his autobiography, "Ce qui me
frappait le plus dans les travaux français, c'était leur
profonde vérité et le soin qu'ils mettaient à écarter les
explications basées sur des apparences.   C'était tout
l'opposé des travaux allemands. En abusant des déduc-
tions, ils avaient fait perdre à la théorie scientifique la
solidité de la charpente."[1]

The French scientist's chief preoccupation is to
measure the importance of his conclusions as exactly
as possible according to the data of experience.   He
is opposed to general theories, hastily built on frag-
mentary facts and which claim to reconstruct reality.
On the opposite side, we see the fondness of German
science for working hypotheses as a means of research.
In this case, a complete theory, elaborately worked out
in every detail, pictures what reality may be from a
small number of given facts.   This is especially evident in the biological sciences where theories thus
pieced together happen to show very close affinities in
structure, in widely different branches.   Such is the
case in Aug. Weismann's ideas on heredity and in those
of Ehrlich on immunity.   One could not overlook this
contrast in the scientific mentality of the two nations.

Another profound characteristic of French science,
as well as of all our national life, is the overwhelming
importance of the rôle played by Paris.   This is related
to the country's unification since the seventeenth century, when everything gravitated around the seat of
royalty.   Centralization was achieved all the better in
the scientific domain, since there was no university life
directed toward the sciences.   Outside of Montpellier,

---

[1] "What used to strike me most in the work of the French was its
profound truth, and the care they took to reject explanations based
on appearances.   It was the exact opposite of the German method.
By abusing the use of deduction, it had caused scientific theory
to be deprived of a solid framework."

the scientific movement had no center in the provinces. The institutions which harbored scientific development during the seventeenth and eighteenth centuries appeared, therefore, in Paris—the *Académie* (the academies formed in the provinces were purely literary societies), the Observatory, the special schools, (except the one at Mézières, which did not have its roots in the place where it stood, but drew its energy from Paris). In this way, Paris has become a center, where all ideas and all intellectual activities converge, and up to a very recent period no truly scientific life could possibly develop in the face of it. Rare are the great careers in science which have been shaped outside of Paris. Paris has absorbed all that appeared outside its walls. This has had an undeniable disadvantage in that it made difficult the development of any ideas which contradicted those of the eminent men who enjoyed unlimited authority at the moment.

These various factors must be taken into account in judging the evolution of French science. Moreover, there is no system which has not both advantages and disadvantages. We must judge the total system according to the first. French science must be estimated according to the important ideas and discoveries it has stirred to life, and by the great men for whose growth it has been responsible. In these respects, French science can bear comparison with that of any country whatever.

There is a last phase which can inspire only sympathy and respect: French science has always been deeply imbued with generous and humane preoccupations. The men who have given it its highest distinction have seen in science not an instrument of power and domination, but a means of promoting and bettering man's state. There is nothing less imperialistic than the spirit of French science, nothing more humanitarian. You will see this in two ideas, expressed a century apart, under very different conditions.

A few months before his death, in 1793, Lavoisier wrote: "Il n'est pas indispensable, pour bien mériter de l'humanité et pour payer son tribut à la patrie, d'être appelé aux fonctions publiques qui concourent à l'organisation et à la génération des empires. Le physicien peut aussi, dans le silence du laboratoire, exercer des fonctions patriotiques. Il peut espérer, par ses travaux diminuer la somme des maux qui affligent l'espèce humaine, augmenter ses jouissances et son bonheur et aspirer au titre glorieux de bienfaiteur de l'humanité."[1]

In 1888, I had, myself, the privilege of hearing Pasteur, at the inauguration of the institute which bears his name. Addressing himself to the Head of the State, who, at that time, was Sadi Carnot's nephew and had exactly the same name, Pasteur made the following statements whose significance has been brought into bolder relief by subsequent events: "S'il m'était permis de terminer par une réflexion philosophique, provoquée en moi par votre présence dans cette salle de travail, je dirais que deux lois contraires semblent aujourd'hui en lutte: une loi de sang et de mort qui, en imaginant chaque jour de nouveaux moyens de combat oblige les peuples à etre toujours prêts pour le champ de bataille, et une loi de paix, de travail, de salut qui ne songe qu'à délivrer l'homme des fléaux qui l'assiègent. L'une ne cherche que les conquêtes violentes, l'autre que le soulagement de l'humanité. Celle-ci met une vie humaine au-dessus de toutes les

---

[1] "It is not indispensable, in order to distinguish oneself in the service of humanity, or to pay one's debt to one's country, to be called to public offices which act in the organization and creation of empires. The physicist also can perform patriotic tasks in the silence of his laboratory. He can hope to diminish, by his work, the mass of ills that afflict the human species, to increase its pleasures and its happiness, and to aspire to the glorious title of benefactor of humanity."

victoires; celle-là sacrifierait volontiers des centaines de mille existences à l'ambition d'un seul. Laquelle de ces deux lois l'emportera sur l'autre, Dieu seul le sait. Mais ce que nous pouvons assurer, c'est que la science française se sera efforcée, en obéissant à la loi d'humanité, de reculer les frontières de la Vie."[1]

In 1892, at the time of his jubilee—the last public function at which he spoke—he came back to the same idea. And it is with this idea that I should like to conclude these talks, as it is even now, and perhaps now more than ever, the true expression of French thought. "Vous, délégués des nations étrangères," he said, "qui êtes venus de si loin, donner une preuve de sympathie à la France, vous m'apportez la joie la plus profonde que puisse éprouver un homme qui croit invinciblement que la science et la paix triompheront de l'ignorance et de la guerre, que les peuples s'uniront, non pour détruire, mais pour édifier et que l'avenir appartiendra à ceux qui auront fait le plus pour l'humanité souffrante."[2]

Such has been in the past, and still is, the French ideal, the ideal of the scientist and that of the whole nation.

[1] "If I were permitted to conclude with a philosophical thought, awakened in me by your presence in this workroom, I should say that two antagonistic laws seem to be battling today; one, a law of blood and death, which, by constantly devising new methods of war, forces people to be constantly prepared for the battle-field; the other, a law of peace, of work and welfare, which thinks only of delivering man from the scourges which beset him. One seeks bloody conquests only, the other only the relief of humanity. This one places a single human life above all victories; that one would sacrifice hundreds of thousands of lives to the ambition of a single man. Which of these laws will triumph, God only knows. But we can affirm that French science will have made every effort to extend the boundaries of Life, by obeying the humane law."

[2] "Delegates of foreign countries," he said, "you who have come from so far away to give France a proof of your friendship, you bring me the deepest joy that can possibly be felt by a man who staunchly believes that science and peace will triumph over ignorance and war, that the nations will unite, not to destroy, but to build, and that the future will belong to those who will have done the most for suffering humanity."

BIBLIOGRAPHY

A bibliography properly stated of the contents of this book would logically include the works (at least the principal ones) of the authors whose names appear in the index on the last pages. It is obviously impossible to give such a bibliography here. A certain number of important references appear as footnotes throughout the various chapters.

We will consequently limit ourselves to point out works or categories of works which are interesting to consult from an historical point of view.

1°—A summary bibliography of the principal works of French scientists appears as an appendix to the various chapters of *La Science Française* (1ère édition 1915, 2ème édition 1933).

2°—The Notices on the *Titres et Travaux scientifiques,* which it is customary to draw up in connection with a candidacy for a chair in the important scientific institutions or for the Academy of Science, constitute a most important bibliographical source for the last century.

3°—Of great importance also are the *Leçons inaugurales* given at the time of assuming a chair and which are generally devoted to a study of the work of the immediate predecessor or of several of the predecessors in the chair. A great number of these *Leçons inaugurales* have been published, notably in the *Revue Scientifique* and in the *Revue générale des Sciences pures et appliquées.*

4°—Another category of studies which present a great interest is that formed by the *Éloges académiques,* notably those pronounced by the *Secrétaires perpétuels* of the Academy of Science and published in the *Mémoires* of the Academy. A certain number have been assembled in volumes:

ARAGO (F.) : *Notices biographiques.* Œuvres complètes d'Arago, t. I-III, 1854.

BERTRAND (JOS.) : *Éloges académiques,* 1 vol., 1880. *Nouvelle série,* 1 vol., 1902.

BERTHELOT (MARCELIN): *Science et éducation et notices académiques*, 1 vol., 1901.

CUVIER (G.): *Recueil des éloges historiques des membres de l'Académie royale des Sciences*, 3 vol., 1819-1827.
—*Éloges historiques* (avec celui l'auteur par P. Flourens), s. d.

DARBOUX (G.): *Éloges académiques*, 1 vol., 1912.

DUMAS (J.-B.): *Discours et éloges académiques*, 1 vol., 1885.

FLOURENS (P.): *Éloges historiques*, 2 vol., 1856.

LACROIX (ALF.): *Figure de savants*, 2 vol., 1932.

PICARD (EM.): *Discours et Mélanges*, 1922.— *Mélanges de Mathématiques et de Physique*, 1924. —*Éloges et discours académiques*, 1932.

BERTRAND (JOS.): *L'Académie des Sciences et les Académiciens de 1666 à 1793*, 1 vol., 1869.

MAURY (L. F. A.): *L'ancienne Académie des Sciences*, 1 vol., 1864.

Likewise, consult all the biographies. A certain number, particularly important (Lavoisier, Pasteur, Ampère, Gerhardt, etc.), have been referred to throughout the book. Also might be mentioned:

LANDRIEU (M.): *Lamarck, le fondateur du transformisme, sa vie, son œuvre*, 1 vol., 1909.

GEOFFROY SAINT-HILAIRE (Isid.): *Vie, travaux et doctrine d'Étienne Geoffroy Saint-Hilaire*, 1 vol., 1847.

FLOURENS (P.): *Histoire des travaux et des idées de Buffon*, 1 vol., 1855.
—*Analyse raisonnée des travaux de Cuvier*, 2 vol., 1841.

PICARD (EM.): *L'œuvre de Henri Poincaré.—La vie et l'œuvre de P. Duhem.—La théorie de l'optique et l'œuvre d'Hippolyte Fizeau.*

5°—Reports and commemorative publications, centenary exercises of scientific societies or institutions, etc.
An especially important series of reports of this kind were published for *l'Exposition universelle de 1867:* notably,

BERNARD (CLAUDE): *Rapport sur les progrès et la marche de la Physiologie générale en France.*

BRONGNIART (A.-D.): *Rapport sur les progrès de la Botanique phytographique.*

DUCHARTRE (P.): *Rapport sur les progrès de la Botanique physiologique.*

MILNE-EDWARDS (H.): *Rapport sur les progrès des sciences zoologiques en France.*

CUVIER (G.): *Histoire des progrès des sciences naturelles depuis 1789 jusqu'à nos jours* (1830), 5 vol. (1828-1836).

ARCHIAC (d'): *Histoire des progrès de la Géologie de 1834 à 1859,* 8 vol., 1847-1860.

GOSSELET (J.): *Constant Prévost, coup d'œil rétrospectif sur la Géologie en France pendant la première moitié du XIXᵉ siècle,* 1896.

SAINTE-CLAIRE DEVILLE (CH.): *Coup d'œil historique sur la Géologie et les travaux d'Élie de Beaumont,* 1878.

LAPPARENT (A. de): *Rapport d'ensemble sur les travaux de la Société géologique de France depuis sa fondation.* 1880. (Bull. Soc. Géol. France, sér. 3, t. VIII).

MARGERIE (EMM. de): *La Société géologique de France de 1880 à 1929* (Livre jubilaire publié à l'occasion du Centenaire de la Société, 1930).

GLEY (EM.): *Essai de Philosophie et d'Histoire de la Biologie,* 1900 (composé pour le cinquantenaire de la Société de Biologie, C. R. Soc. Biologie, t. LI, 1898).

PEYERIMHOFF (P. de): *La Société entomologique de France* (1832-1931). Livre du Centenaire de la Soc. Entom. de France, 1932.

—*Le Collège de France* (1530-1930). Livre jubilaire compose à l'occasion de son quatrième centenaire, 1932.

—*Centenaire de l'École normale supérieure,* 1895.

—*Centenaire du Muséum d'Histoire Naturelle* (vol. commémoratif), 1893.[1]

CAP (P. A.): *Le Muséum d'Histoire Naturelle. Histoire de sa fondation et des développements successifs de l'établissement,* 1843.

DELEUZE (J. P. E.): *Histoire et description du Muséum Royal d'Histoire naturelle,* etc., 2 vol., 1823.

—*Un demi-siècle de civilisation française* (1870-1915). Paris, 1916, avec des chapitres sur la Chimie (G. Lemoine), les Sciences naturelles (Ed. Perrier), les Sciences mathématiques (Em. Picard), la Physique (L. Poincaré), les Sciences biologiques et médicales (Ch. Richet).

OCAGNE (MAURICE d'): *Les Sciences en France depuis* 1870 (La Troisième République), 1933.

—*Hommes et choses de sciences,* 2 vol. 1930.

6°—Miscellaneous works.

PICARD (EM.): *La science moderne et son état actuel,* 1906.

WURTZ (AD.): *Histoire des doctrines chimiques depuis Lavoisier* (Dictionnaire de chimie pure et appliquée, T. I., Discours préliminaire, 1874).

QUATREFAGES (A. de): *Charles Darwin et ses précurseurs français,* 2 vol., Paris, 189—.

PERRIER (ED.): *La Philosophie zoologique avant Darwin,* 1884.

ROSTAND (J.): *La formation de l'être; Histoire des idées sur la génération,* 1930.

---

[1] E. T. Hamy published numerous historical studies and documents on the subject of this institution. Moreover, it is customary to publish, at the death of the professors or former professors, a list of their works in the *Archives du Muséum.*

7°—A few works on the history of the sciences:

LA SCIENCE FRANÇAISE, 2 vol., 1re éd., 1915, 2e éd., 1933.

HISTOIRE DE LA NATION FRANÇAISE, par G. HANOTAUX, T. XIV et XV. Histoire des sciences en France:

ANDOYER (H.) et HUMBERT (P.): *Histoire des sciences mathématiques,* T. XIV, 1924.

FABRY (CH.): *Histoire de la Physique,* Ibid.

COLSON (CL.): *Histoire de la Chimie,* Ibid.

CAULLERY (M.): *Histoire des sciences biologiques.* T. XV, 1925.

MONTUCLA: *Histoire des mathématiques,* 2 vol., 1758; seconde édition, complétée par Jerôme de Lalande, 4 vol., 1799-1802.

DELAMBRE: *Histoire de l'Astronomie,* 6 vol., 1817-1827.

WOLF (CH.): *Histoire de l'Observatoire de Paris de sa fondation à 1793,* 1902.

MARIE (MAXIMILIEN): *Histoire des sciences mathématiques et physiques,* 8 vol., 1883-1885.

HŒFER (F.): *Histoire de la Physique.*

HŒFER (F.): *Histoire de la Chimie,* 2 vol., 1866-1869.

DELACRE (M.): *Histoire de la Chimie,* 1920.

KIRRMANN: *La Chimie d'hier et d'aujourd'hui,* 1928.

CUVIER (G.): *Histoire des sciences naturelles depuis leurs origines jusqu'à nos jours,* 5 vol., 1846.

DAREMBERG (G.): *Histoire des sciences médicales,* 2 vol., 1870.

CARUS (V.): *Histoire de la Zoologie* (traduction française), 1880.

SACHS (J.): *Histoire de la Botanique* (trad. française), 1892.

COMBES (R.): *Histoire de la Biologie végétale en France,* 1933.

INDEX OF PROPER NAMES

HISTORY, PHILOSOPHY AND
SOCIOLOGY OF SCIENCE

Classics, Staples and Precursors

An Arno Press Collection

Aliotta, [Antonio]. **The Idealistic Reaction Against Science.** 1914

Arago, [Dominique François Jean]. **Historical Eloge of James Watt.** 1839

Bavink, Bernhard. **The Natural Sciences.** 1932

Benjamin, Park. **A History of Electricity.** 1898

Bennett, Jesse Lee. **The Diffusion of Science.** 1942

[Bronfenbrenner], Ornstein, Martha. **The Role of Scientific Societies in the Seventeenth Century.** 1928

Bush, Vannevar. **Endless Horizons.** 1946

Campanella, Thomas. **The Defense of Galileo.** 1937

Carmichael, R. D. **The Logic of Discovery.** 1930

Caullery, Maurice. **French Science and its Principal Discoveries Since the Seventeenth Century.** [1934]

Caullery, Maurice. **Universities and Scientific Life in the United States.** 1922

Debates on the Decline of Science. 1975

de Beer, G. R. **Sir Hans Sloane and the British Museum.** 1953

Dissertations on the Progress of Knowledge. [1824]. 2 vols. in one

Euler, [Leonard]. **Letters of Euler.** 1833. 2 vols. in one

Flint, Robert. **Philosophy as Scientia Scientiarum and a History of Classifications of the Sciences.** 1904

Forke, Alfred. **The World-Conception of the Chinese.** 1925

Frank, Philipp. **Modern Science and its Philosophy.** 1949

The Freedom of Science. 1975

George, William H. **The Scientist in Action.** 1936

Goodfield, G. J. **The Growth of Scientific Physiology.** 1960

Graves, Robert Perceval. **Life of Sir William Rowan Hamilton.** 3 vols. 1882

Haldane, J. B. S. **Science and Everyday Life.** 1940

Hall, Daniel, et al. **The Frustration of Science.** 1935

Halley, Edmond. **Correspondence and Papers of Edmond Halley.** 1932

Jones, Bence. **The Royal Institution.** 1871

Kaplan, Norman. **Science and Society.** 1965

Levy, H. **The Universe of Science.** 1933

Marchant, James. **Alfred Russel Wallace.** 1916

McKie, Douglas and Niels H. de V. Heathcote. **The Discovery of Specific and Latent Heats.** 1935

Montagu, M. F. Ashley. **Studies and Essays in the History of Science and Learning.** [1944]

Morgan, John. **A Discourse Upon the Institution of Medical Schools in America.** 1765

Mottelay, Paul Fleury. **Bibliographical History of Electricity and Magnetism Chronologically Arranged.** 1922

Muir, M. M. Pattison. **A History of Chemical Theories and Laws.** 1907

National Council of American-Soviet Friendship. **Science in Soviet Russia: Papers Presented at Congress of American-Soviet Friendship.** 1944

Needham, Joseph. **A History of Embryology.** 1959

Needham, Joseph and Walter Pagel. **Background to Modern Science.** 1940

Osborn, Henry Fairfield. **From the Greeks to Darwin.** 1929

Partington, J[ames] R[iddick]. **Origins and Development of Applied Chemistry.** 1935

Polanyi, M[ichael]. **The Contempt of Freedom.** 1940

Priestley, Joseph. **Disquisitions Relating to Matter and Spirit.** 1777

Ray, John. **The Correspondence of John Ray.** 1848

Richet, Charles. **The Natural History of a Savant.** 1927

Schuster, Arthur. **The Progress of Physics During 33 Years (1875-1908).** 1911

Science, Internationalism and War. 1975

Selye, Hans. **From Dream to Discovery: On Being a Scientist.** 1964

Singer, Charles. **Studies in the History and Method of Science.** 1917/1921. 2 vols. in one

Smith, Edward. **The Life of Sir Joseph Banks.** 1911

Snow, A. J. **Matter and Gravity in Newton's Physical Philosophy.** 1926

Somerville, Mary. **On the Connexion of the Physical Sciences.** 1846

Thomson, J. J. **Recollections and Reflections.** 1936

Thomson, Thomas. **The History of Chemistry.** 1830/31

Underwood, E. Ashworth. **Science, Medicine and History.** 2 vols. 1953

Visher, Stephen Sargent. **Scientists Starred 1903-1943 in American Men of Science.** 1947

Von Humboldt, Alexander. **Views of Nature: Or Contemplations on the Sublime Phenomena of Creation.** 1850

Von Meyer, Ernst. **A History of Chemistry from Earliest Times to the Present Day.** 1891

Walker, Helen M. **Studies in the History of Statistical Method.** 1929

Watson, David Lindsay. **Scientists Are Human.** 1938

Weld, Charles Richard. **A History of the Royal Society.** 1848. 2 vols. in one

Wilson, George. **The Life of the Honorable Henry Cavendish.** 1851